FERMAT'S LAST
THEOREM

FERMAT'S LAST THEOREM

The story of a riddle that confounded the
world's greatest minds for 358 years

SIMON SINGH

FOURTH ESTATE • *London*

First published in Great Britain in 1997 by
Fourth Estate Limited
6 Salem Road
London W2 4BU

7 9 10 8

A catalogue record for this book is available from the
British Library.

ISBN 1-85702-521-0

Typeset by MATS, Southend-on-Sea, Essex
Printed in Great Britain by Clays Ltd, St Ives, plc

In memory
of
Pakhar Singh Birring

CONTENTS

FOREWORD

We finally met across a room, not crowded, but large enough to hold the entire Mathematics Department at Princeton on their occasions of great celebration. On that particular afternoon, there were not so very many people around, but enough for me to be uncertain as to which one was Andrew Wiles. After a few moments I picked out a shy-looking man, listening to the conversation around him, sipping tea, and indulging in the ritual gathering of minds that mathematicians the world over engage in at around four o'clock in the afternoon. He simply guessed who I was.

It was the end of an extraordinary week. I had met some of the finest mathematicians alive, and begun to gain an insight into their world. But despite every attempt to pin down Andrew Wiles, to speak to him, and to convince him to take part in a BBC *Horizon* documentary film on his achievement, this was our first meeting. This was the man who had recently announced that he had found the holy grail of mathematics; the man who claimed he had proved Fermat's Last Theorem. As we spoke, Wiles had a distracted and withdrawn air about him, and although he was polite and friendly, it was clear that he wished me as far away from him as possible. He explained very simply that he could not possibly focus on anything but his work, which was at a critical stage, but perhaps later, when the current pressures had been resolved, he would be pleased to take part. I knew, and he knew I knew, that he was facing the collapse of his life's ambition, and that the holy grail he had held was now being revealed as no more than a rather beautiful, valuable, but straightforward drinking vessel. He had found a flaw in his heralded proof.

The story of Fermat's Last Theorem is unique. By the time I first met Andrew Wiles, I had come to realise that it is truly one of the greatest stories in the sphere of scientific or academic endeavour. I had seen the headlines in the summer of 1993, when the proof had put maths on the front pages of national newspapers around the world. At that time I had only a vague recollection of what the Last Theorem was, but saw that it was obviously something very special, and something that had the smell of a *Horizon* film to it. I spent the next weeks talking to many mathematicians: those closely involved in the story, or close to Andrew, and those who simply shared the thrill of witnessing a great moment in their field. All generously shared their insights into mathematical history, and patiently talked me through what little understanding I could achieve of the concepts involved. Rapidly it became clear that this was subject matter that perhaps only half a dozen people in the world could fully grasp. For a while I wondered if I was insane to attempt to make a film. But from those mathematicians I also learned of the rich history, and the deeper significance of Fermat to mathematics and its practitioners, and that, I realized, was where the real story lay.

I learned of the ancient Greek origins of the problem, and that Fermat's Last Theorem was the Himalayan peak of number theory. I was introduced to the aesthetic beauty of maths, and I began to appreciate what it is to describe mathematics as the language of nature. Through Wiles's contemporaries I grasped the herculean nature of his work in pulling together all the most recent techniques of number theory to apply to his proof. From his friends in Princeton I heard of the intricate progress of Andrew's years of isolated study. I built up an extraordinary picture around Andrew Wiles, and the puzzle that dominated his life, but I seemed destined never to meet the man himself.

Although the maths involved in Wiles's proof is some of the

toughest in the world, I found that the beauty of Fermat's Last Theorem lies in the fact that the problem itself is supremely simple to understand. It is a puzzle that is stated in terms familiar to every schoolchild. Pierre de Fermat was a man in the Renaissance tradition, who was at the centre of the rediscovery of ancient Greek knowledge, but he asked a question that the Greeks would not have thought to ask, and in so doing produced what became the hardest problem on earth for others to solve. Tantalisingly, he left a note for posterity suggesting that he had an answer, but not what it was. That was the beginning of the chase that lasted three centuries.

That time-span underlies the significance of this puzzle. It is hard to conceive of any problem, in any discipline of science, so simply and clearly stated that could have withstood the test of advancing knowledge for so long. Consider the leaps in understanding in physics, chemistry, biology, medicine and engineering that have occurred since the seventeenth century. We have progressed from 'humours' in medicine to gene-splicing, we have identified the fundamental atomic particles, and we have placed men on the moon, but in number theory Fermat's Last Theorem remained inviolate.

For some time in my research I looked for a reason why the Last Theorem mattered to anyone but a mathematician, and why it would be important to make a programme about it. Maths has a multitude of practical applications, but in the case of number theory the most exciting uses that I was offered were in cryptography, in the design of acoustic baffling, and in communication from distant spacecraft. None of these seemed likely to draw in an audience. What was far more compelling were the mathematicians themselves, and the sense of passion that they all expressed when talking of Fermat.

Maths is one of the purest forms of thought, and to outsiders mathematicians may seem almost other-worldly. The thing that struck me in all my discussions with them was the extraordinary precision of their conversation. A question was rarely answered immediately, I would often have to wait while the precise structure of the answer was resolved in the mind, but it would then emerge, as articulate and careful a statement as I could have wished for. When I tackled Andrew's friend Peter Sarnak on this, he explained that mathematicians simply hate to make a false statement. Of course they use intuition and inspiration, but formal statements have to be absolute. Proof is what lies at the heart of maths, and is what marks it out from other sciences. Other sciences have hypotheses that are tested against experimental evidence until they fail, and are overtaken by new hypotheses. In maths, absolute proof is the goal, and once something is proved, it is proved forever, with no room for change. In the Last Theorem, mathematicians had their greatest challenge of proof, and the person who found the answer would receive the adulation of the entire discipline.

Prizes were offered, and rivalry flourished. The Last Theorem has a rich history that touches death and deception, and it has even spurred on the development of maths. As the Harvard mathematician Barry Mazur has put it, Fermat added a certain 'animus' to those areas of maths that were associated with early attempts at the proof. Ironically, it turned out that just such an area of maths was central to Wiles's final proof.

Gradually picking up an understanding of this unfamiliar field, I came to appreciate Fermat's Last Theorem as central to, and even a parallel for the development of maths itself. Fermat was the father of modern number theory, and since his time mathematics had evolved, progressed and diversified into many arcane areas, where new techniques had spawned new areas of maths, and become ends in themselves. As the centuries passed, the Last Theorem came to seem less and less relevant to the cutting edge of mathematical research, and more and more turned into a curiosity. But it is now clear that its centrality to maths never diminished.

Problems around numbers, such as the one Fermat posed, are like playground puzzles, and mathematicians like solving puzzles. To Andrew Wiles it was a very special puzzle, and nothing less than his life's ambition. Thirty years before, as a child, he had been inspired by Fermat's Last Theorem, having stumbled upon it in a public library book. His childhood and adulthood dream was to solve the problem, and when he first revealed a proof in that summer of 1993, it came at the end of seven years of dedicated work on the problem, a degree of focus and determination that is hard to imagine. Many of the techniques he used had not been created when he began. He also drew together the work of many fine mathematicians, linking ideas and creating concepts that others had feared to attempt. In a sense, reflected Barry Mazur, it turned out that everyone had been working on Fermat, but separately and without having it as a goal, for the proof had required all the power of modern maths to be brought to bear upon its solution. What Andrew had done was tie together once again areas of maths that had seemed far apart. His work therefore seemed to be a justification of all the diversification that maths had undergone since the problem had been stated.

At the heart of his proof of Fermat, Andrew had proved an idea known as the Taniyama–Shimura Conjecture, which created a new bridge between wildly different mathematical worlds. For many, the goal of one unified mathematics is supreme, and this was a glimpse of just such a world. So in proving Fermat, Andrew Wiles had cemented some of the most important number theory of the post-war period, and had secured the base of a pyramid of conjectures that were built upon it. This was no longer simply solving the longest-standing mathematical puzzle, but was pushing the very boundaries of mathematics itself. It was as if Fermat's simple problem, born at a time when maths was in its infancy, had been waiting for this moment.

The story of Fermat had ended in the most spectacular fashion. For Andrew Wiles, it meant the end of professional isolation of a kind almost alien to maths, which is usually a collaborative activity. Ritual afternoon tea in mathematics institutes the world over is a time when ideas come together, and sharing insight before publication is the norm. Ken Ribet, a mathematician who was himself central to the proof, only half jokingly suggested to me that it is the insecurity of mathematicians that requires the support structure of their colleagues. Andrew Wiles had eschewed all that, and kept his work to himself in all but the final stages. That too was a measure of the importance of Fermat. He had a real driving passion to be the one who solved this problem, a passion strong enough to devote seven years of his life and keep his goal to himself. He had known that however irrelevant the problem had seemed, competition for Fermat had never lessened, and he could never have risked revealing what he was doing.

After weeks of researching the field, I had arrived in Princeton. For mathematicians, the level of emotion was intense. I had found a story of competition, success, isolation, genius, triumph, jealousy, intense pressure, loss and even tragedy. At the heart of that crucial Taniyama–Shimura Conjecture lay the tragic post-war life in Japan of Yutaka Taniyama, whose story I was privileged to hear from his close friend Goro Shimura. From Shimura too I learned of the notion of 'goodness' in maths, where things simply feel right, because they are good. Somehow, the sense of goodness pervaded the atmosphere of mathematics that summer. All were revelling in the glorious moment.

With all this in train, small wonder at the weight of responsibility that Andrew felt as the flaw had gradually emerged over the autumn of 1993. With the eyes of the world upon him, and his colleagues calling to have the proof made public, somehow, and only he knows how, he didn't crack. He had moved from doing maths in privacy and at his own pace to suddenly working in public. Andrew is an intensely private man, who fought hard to keep his family sheltered from the storm that was breaking around him. Throughout that week while I was in Princeton, I called, I left notes at his office, on his doorstep, and with his friends; I even provided a gift of English tea and Marmite. But he resisted my overtures, until that chance meeting on the day of my departure. A quiet, intense conversation followed, that in the end lasted barely fifteen minutes.

When we parted that afternoon there was an understanding between us. If he managed to repair the proof, then he would come to me to discuss a film; I was prepared to wait. But as I flew home to London that night it seemed to me that the television programme was dead. No one had ever repaired a hole in the many attempted proofs of Fermat in three centuries. History was littered with false claims, and much as I wished that he would be the exception, it was hard to imagine Andrew as anything but another headstone in that mathematical graveyard.

A year later I received the call. After an extraordinary mathematical twist, and a flash of true insight and inspiration, Andrew had finally brought an end to Fermat in his professional life. A year after that, we found the time for him to devote to filming. By this time I had invited Simon Singh to join me in making the film, and together we spent time with Andrew, learning from the man himself the full story of those seven years of isolated study, and his year of hell that followed. As we filmed, Andrew told us, as he had told no one before, of his innermost feelings about what he had done; how for thirty years he had hung on to a childhood dream; how so much of the maths he had ever studied had been, without his really knowing it at the time, really a gathering of tools for the Fermat challenge that had dominated his career; how nothing would ever be the same; of his sense of loss for the problem that would no longer be his constant companion; and of the uplifting sense of release that he now felt. For a field in which the subject matter is technically about as difficult for a lay audience to understand as can be imagined, the level of emotional charge in our conversations was greater than any I have experienced in a career in science film making. For Andrew it was the end of a chapter in his life. For me it was a privilege to be close to it.

The film was transmitted on BBC Television as *Horizon: Fermat's Last Theorem*. Simon Singh has now developed those insights and intimate conversations, together with the full richness of the Fermat story and the history and mathematics that have always hung around it, into this book, which is a complete and enlightening record of one of the greatest stories in human thinking.

John Lynch
Editor of BBC TV's *Horizon* series
March 1997

PREFACE

The story of Fermat's Last Theorem is inextricably linked with the history of mathematics, touching on all the major themes of number theory. It provides a unique insight into what drives mathematics and, perhaps more importantly, what inspires mathematicians. The Last Theorem is at the heart of an intriguing saga of courage, skulduggery, cunning and tragedy, involving all the greatest heroes of mathematics.

Fermat's Last Theorem has its origins in the mathematics of ancient Greece, two thousand years before Pierre de Fermat constructed the problem in the form we know it today. Hence, it links the foundations of mathematics created by Pythagoras to the most sophisticated ideas in modern mathematics. In writing this book I have chosen a largely chronological structure which begins by describing the revolutionary ethos of the Pythagorean Brotherhood, and ends with Andrew Wiles's personal story of his struggle to find a solution to Fermat's conundrum.

Chapter 1 tells the story of Pythagoras, and describes how Pythagoras' theorem is the direct ancestor of the Last Theorem. This chapter also discusses some of the fundamental concepts of mathematics which will recur throughout the book. Chapter 2 takes the story from ancient Greece to seventeenth-century France, where Pierre de Fermat created the most profound riddle in the history of mathematics. To convey the extraordinary character of Fermat and his contribution to mathematics, which goes far beyond the Last Theorem, I have spent several pages describing his life, and some of his other brilliant discoveries.

Chapters 3 and 4 describe some of the attempts to prove Fermat's Last Theorem during the eighteenth, nineteenth and early twentieth centuries. Although these efforts ended in failure they led to a marvellous arsenal of mathematical techniques and tools, some of which have been integral to the very latest attempts to prove the Last Theorem. In addition to describing the mathematics I have devoted much of these chapters to the mathematicians who became obsessed by Fermat's legacy. Their stories show how mathematicians were prepared to sacrifice everything in the search for truth, and how mathematics has evolved through the centuries.

The remaining chapters of the book chronicle the remarkable events of the last forty years which have revolutionised the study of Fermat's Last Theorem. In particular Chapters 6 and 7 focus on the work of Andrew Wiles, whose breakthroughs in the last decade astonished the mathematical community. These later chapters are based on extensive interviews with Wiles. This was a unique opportunity for me to hear at first hand one of the most extraordinary intellectual journeys of the twentieth century and I hope that I have been able to convey the creativity and heroism that was required during Wiles's ten-year ordeal.

In telling the tale of Pierre de Fermat and his baffling riddle I have tried to describe the mathematical concepts without resorting to equations, but inevitably x, y and z do occasionally rear their ugly heads. When equations do appear in the text I have endeavoured to provide sufficient explanation such that even readers with no background in mathematics will be able to understand their significance. For those readers with a slightly deeper knowledge of the subject I have provided a series of appendices which expand on the mathematical ideas contained in the main text. In addition I have included a list of further reading, which is generally aimed at providing the layperson with more detail about particular areas of mathematics.

This book would not have been possible without the help and involvement of many people. In particular I would like to thank Andrew Wiles, who went out of his way to give long and detailed interviews during a time of intense pressure. During my seven years as a science journalist I have never met anybody with a greater level of passion and commitment to their subject, and I am eternally grateful that Professor Wiles was prepared to share his story with me.

I would also like to thank the other mathematicians who helped me in the writing of this book and who allowed me to interview them at length. Some of them have been deeply involved in tackling Fermat's Last Theorem, while others were witnesses to the historic events of the last forty years. The hours I spent quizzing and chatting with them were enormously enjoyable and I appreciate their patience and enthusiam while explaining so many beautiful mathematical concepts to me. In particular I would like to thank John Coates, John Conway, Nick Katz, Barry Mazur, Ken Ribet, Peter Sarnak, Goro Shimura and Richard Taylor.

I have tried to illustrate this book with as many portraits as possible to give the reader a better sense of the characters involved in the story of Fermat's Last Theorem. Various libraries and archives have gone out of their way to help me, and in particular I would like to thank Susan Oakes of the London Mathematical Society, Sandra Cumming of the Royal Society and Ian Stewart of Warwick University. I am also grateful to Jacquelyn Savani of Princeton University, Duncan McAngus, Jeremy Gray, Paul Balister and the Isaac Newton Institute for their help in finding research material. Thanks also go to Patrick Walsh, Christopher Potter, Bernadette Alves, Sanjida O'Connell and my parents for their comments and support throughout the last year.

Finally, many of the interviews quoted in this book were obtained while I was working on a television documentary on the subject of Fermat's Last Theorem. I would like to thank the BBC for allowing me to use this material, and in particular I owe a debt of gratitude to John Lynch, who worked with me on the documentary, and who helped to inspire my interest in the subject.

Simon Singh
Thakarki, Phagwara
1997

FERMAT'S LAST
THEOREM

Andrew Wiles aged ten years, when he first encountered Fermat's Last Theorem.

1

'I Think I'll Stop Here'

Archimedes will be remembered when Aeschylus is forgotten, because languages die and mathematical ideas do not. 'Immortality' may be a silly word, but probably a mathematician has the best chance of whatever it may mean.

G.H. Hardy

23 June 1993, Cambridge

It was the most important mathematics lecture of the century. Two hundred mathematicians were transfixed. Only a quarter of them fully understood the dense mixture of Greek symbols and algebra that covered the blackboard. The rest were there merely to witness what they hoped would be a truly historic occasion.

The rumours had started the previous day. Electronic mail over the Internet had hinted that the lecture would culminate in a solution to Fermat's Last Theorem, the world's most famous mathematical problem. Such gossip was not uncommon. The subject of Fermat's Last Theorem would often crop up over tea, and mathematicians would speculate as to who might be doing what. Sometimes mathematical mutterings in the senior common room would turn the speculation into rumours of a breakthrough, but nothing had ever materialised.

This time the rumour was different. One Cambridge research

1

student was so convinced that it was true that he dashed to the bookies to bet £10 that Fermat's Last Theorem would be solved within the week. However, the bookie smelt a rat and refused to accept his wager. This was the fifth student to have approached him that day, all of them asking to place the identical bet. Fermat's Last Theorem had baffled the greatest minds on the planet for over three centuries, but now even bookmakers were beginning to suspect that it was on the verge of being proved.

The three blackboards became filled with calculations and the lecturer paused. The first board was erased and the algebra continued. Each line of mathematics appeared to be one tiny step closer to the solution, but after thirty minutes the lecturer had still not announced the proof. The professors crammed into the front rows waited eagerly for the conclusion. The students standing at the back looked to their seniors for hints of what the conclusion might be. Were they watching a complete proof to Fermat's Last Theorem, or was the lecturer merely outlining an incomplete and anticlimactic argument?

The lecturer was Andrew Wiles, a reserved Englishman who had emigrated to America in the 1980s and taken up a professorship at Princeton University where he had earned a reputation as one of the most talented mathematicians of his generation. However, in recent years he had almost vanished from the annual round of conferences and seminars, and colleagues had begun to assume that Wiles was finished. It is not unusual for brilliant young minds to burn out, a point noted by the mathematician Alfred Adler: 'The mathematical life of a mathematician is short. Work rarely improves after the age of twenty-five or thirty. If little has been accomplished by then, little will ever be accomplished.'

'Young men should prove theorems, old men should write books,' observed G.H. Hardy in his book *A Mathematician's Apology*.

'No mathematician should ever forget that mathematics, more than any other art or science, is a young man's game. To take a simple illustration, the average age of election to the Royal Society is lowest in mathematics.' His own most brilliant student Srinivasa Ramanujan was elected a Fellow of the Royal Society at the age of just thirty-one, having made a series of outstanding breakthroughs during his youth. Despite having received very little formal education in his home village of Kumbakonam in South India, Ramanujan was able to create theorems and solutions which had evaded mathematicians in the West. In mathematics the experience that comes with age seems less important than the intuition and daring of youth. When he posted his results to Hardy, the Cambridge professor was so impressed that he invited him to abandon his job as a lowly clerk in South India and attend Trinity College, where he could interact with some of the world's foremost number theorists. Sadly the harsh East Anglian winters were too much for Ramanujan who contracted tuberculosis and died at the age of thirty-three.

Other mathematicians have had equally brilliant but short careers. The nineteenth-century Norwegian Niels Henrik Abel made his greatest contribution to mathematics at the age of nineteen and died in poverty, just eight years later, also of tuberculosis. Charles Hermite said of him, 'He has left mathematicians something to keep them busy for five hundred years', and it is certainly true that Abel's discoveries still have a profound influence on today's number theorists. Abel's equally gifted contemporary Evariste Galois also made his breakthroughs while still a teenager and then died aged just twenty-one.

These examples are not intended to show that mathematicians die prematurely and tragically but rather that their most profound ideas are generally conceived while they are young, and as Hardy once said, 'I do not know an instance of a major mathematical advance initiated by a man past fifty.' Middle-aged mathematicians often fade into the background and occupy their remaining years teaching or administrating rather than researching. In the case of Andrew Wiles nothing could be further from the truth. Although he had reached the grand old age of forty he had spent the last seven years working in complete secrecy, attempting to solve the single greatest problem in mathematics. While others suspected he had dried up, Wiles was making fantastic progress, inventing new techniques and tools which he was now ready to reveal. His decision to work in absolute isolation was a high-risk strategy and one which was unheard of in the world of mathematics.

Without inventions to patent, the mathematics department of any university is the least secretive of all. The community prides itself in an open and free exchange of ideas and tea-time breaks have evolved into daily rituals during which concepts are shared and explored over biscuits and Earl Grey. As a result it is increasingly common to find papers being published by co-authors or teams of mathematicians and consequently the glory is shared out equally. However, if Professor Wiles had genuinely discovered a complete and accurate proof of Fermat's Last Theorem, then the most wanted prize in mathematics was his and his alone. The price he had to pay for his secrecy was that he had not previously discussed or tested any of his ideas with the mathematics community and therefore there was a significant chance that he had made some fundamental error.

Ideally Wiles had wanted to spend more time going over his work to allow him to check fully his final manuscript. Then the unique opportunity arose to announce his discovery at the Isaac Newton Institute in Cambridge and he abandoned caution. The sole aim of the institute's existence is to bring together the world's greatest intellects for a few weeks in order to hold seminars on a cutting-edge research topic of their choice. Situated on the outskirts of the university, away from students and other distractions, the building is especially designed to encourage the academics to concentrate on collaboration and brainstorming. There are no dead-end corridors in which to hide and every office faces a central forum. The mathematicians are supposed to spend time in this open area, and are discouraged from keeping their office doors closed. Collaboration while moving around the institute is also encouraged – even the elevator, which only travels three floors, contains a blackboard. In fact every room in the building has at least one blackboard, including the bathrooms. On this occasion the seminars at the Newton Institute came under the heading of '*L*-functions and Arithmetic'. All the world's top number theorists had been gathered together in order to discuss problems relating to this highly specialised area of pure mathematics, but only Wiles realised that *L*-functions might hold the key to solving Fermat's Last Theorem.

Although he had been attracted by having the opportunity to reveal his work to such an eminent audience, the main reason for making the announcement at the Newton Institute was that it was in his home town, Cambridge. This was where Wiles had been born, it was here he grew up and developed his passion for numbers, and it was in Cambridge that he had alighted on the problem which was to dominate the rest of his life.

The Last Problem

In 1963, when he was ten years old, Andrew Wiles was already fascinated by mathematics. 'I loved doing the problems in school, I'd take them home and make up new ones of my own. But the best problem I ever found I discovered in my local library.'

One day, while wandering home from school, young Wiles decided to visit the library in Milton Road. It was rather impoverished compared with the libraries of the colleges, but nonetheless it had a generous collection of puzzle books and this is what often caught Andrew's attention. These books were packed with all sorts of scientific conundrums and mathematical riddles, and for each question the solution would be conveniently laid out somewhere in the final few pages. But this time Andrew was drawn to a book with only one problem, and no solution.

The book was *The Last Problem* by Eric Temple Bell, the history of a mathematical problem which has its roots in ancient Greece, but which only reached full maturity in the seventeenth century. It was then that the great French mathematician Pierre de Fermat inadvertently set it as a challenge for the rest of the world. One great mathematician after another had been humbled by Fermat's legacy and for three hundred years nobody had been able to solve it. There are other unsolved questions in mathematics, but what makes Fermat's problem so special is its deceptive simplicity. Thirty years after first reading Bell's account, Wiles told me how he felt the moment he was introduced to Fermat's Last Theorem: 'It looked so simple, and yet all the great mathematicians in history couldn't solve it. Here was a problem that I, a ten-year-old, could understand and I knew from that moment that I would never let it go. I had to solve it.'

The problem looks so straightforward because it is based on the one piece of mathematics that everyone can remember – Pythagoras' theorem:

> In a right-angled triangle the square on the hypotenuse is
> equal to the sum of the squares on the other two sides.

As a result of this Pythagorean ditty, the theorem has been scorched into millions if not billions of human brains. It is the fundamental theorem that every innocent schoolchild is forced to learn. But despite the fact that it can be understood by a ten-year-old, Pythagoras' creation was the inspiration for a problem which had thwarted the greatest mathematical minds of history.

Pythagoras of Samos was one of the most influential and yet mysterious figures in mathematics. Because there are no first-hand accounts of his life and work, he is shrouded in myth and legend, making it difficult for historians to separate fact from fiction. What seems certain is that Pythagoras developed the idea of numerical logic and was responsible for the first golden age of mathematics. Thanks to his genius numbers were no longer merely used to count and calculate, but were appreciated in their own right. He studied the properties of particular numbers, the relationships between them and the patterns they formed. He realised that numbers exist independently of the tangible world and therefore their study was untainted by the inaccuracies of perception. This meant he could discover truths which were independent of opinion or prejudice and which were more absolute than any previous knowledge.

Living in the sixth century BC, Pythagoras gained his mathematical skills on his travels throughout the ancient world. Some tales would have us believe that he travelled as far as India and Britain, but what is more certain is that he gathered many mathematical techniques and tools from the Egyptians and Babylonians. Both these ancient peoples had gone beyond the limits of simple counting and were capable of performing complex calculations which enabled them to create sophisticated accounting systems and construct elaborate buildings. Indeed they saw mathematics as merely a tool for solving practical problems; the motivation behind discovering some of the basic rules of geometry was to allow reconstruction of field boundaries which were lost in the annual flooding of the Nile. The word itself, geometry, means 'to measure the earth'.

Pythagoras observed that the Egyptians and Babylonians conducted each calculation in the form of a recipe which could be followed blindly. The recipes, which would have been passed down through the generations, always gave the correct answer and so nobody bothered to question them or explore the logic underlying the equations. What was important for these civilisations was that a calculation worked – why it worked was irrelevant.

After twenty years of travel Pythagoras had assimilated all the mathematical rules in the known world. He set sail for his home island of Samos in the Aegean Sea with the intention of founding a school devoted to the study of philosophy and in particular concerned with research into his newly acquired mathematical rules. He wanted to understand numbers, not merely exploit them. He hoped to find a plentiful supply of free-thinking students who could help him develop radical new philosophies, but during his absence the tyrant Polycrates had turned the once liberal Samos into an intolerant and conservative society. Polycrates invited Pythagoras to join his court, but the philosopher realised that this was only a manoeuvre aimed at silencing him and therefore declined the honour. Instead he left the city in favour of a cave in a remote part of the island, where he could contemplate without fear of persecution.

Pythagoras did not relish his isolation and eventually resorted to bribing a young boy to be his first pupil. The identity of the young boy is uncertain but some historians have suggested that his name was also Pythagoras, and that the student would later gain fame as the first person to suggest that athletes should eat meat to improve their physique. Pythagoras, the teacher, paid his student three oboli for each lesson he attended and noticed that as the weeks passed the boy's initial reluctance to learn was transformed into an enthusiasm for knowledge. To test his pupil Pythagoras pretended that he could no longer afford to pay the student and that the lessons would have to stop, at which point the boy offered to pay for his education rather than have it ended. The pupil had become a disciple. Unfortunately this was Pythagoras' only conversion on Samos. He did temporarily establish a school, known as the Semicircle of Pythagoras, but his views on social reform were unacceptable and the philosopher was forced to flee the colony with his mother and his one and only disciple.

Pythagoras departed for southern Italy, which was then a part of Magna Graecia, and settled in Croton where he was fortunate in finding the ideal patron in Milo, the wealthiest man in Croton and one of the strongest men in history. Although Pythagoras' reputation as the sage of Samos was already spreading across Greece, Milo's fame was even greater. Milo was a man of Herculean proportions who had been champion of the Olympic and Pythian Games a record twelve times. In addition to his athleticism Milo also appreciated and studied philosophy and mathematics. He set aside part of his house and provided Pythagoras with enough room to establish a school. So it was that the most creative mind and the most powerful body formed a partnership.

Secure in his new home Pythagoras founded the Pythagorean Brotherhood – a band of six hundred followers who were capable not only of understanding his teachings, but who could add to them by creating new ideas and proofs. Upon entering the Brotherhood each follower had to donate all their worldly possessions to a common fund and should anybody ever leave they would receive twice the amount they had originally donated and a tombstone would be erected in their memory. The Brotherhood was an egalitarian school and included several sisters. Pythagoras' favourite student was Milo's own daughter, the beautiful Theano, and, despite the difference in their ages, they eventually married.

Soon after founding the Brotherhood, Pythagoras coined the word *philosopher*, and in so doing defined the aims of his school. While attending the Olympic Games, Leon, Prince of Phlius, asked Pythagoras how he would describe himself. Pythagoras replied, 'I am a philosopher,' but Leon had not heard the word before and asked him to explain.

Life, Prince Leon, may well be compared with these public Games for in the vast crowd assembled here some are attracted by the acquisition of gain, others are led on by the hopes and ambitions of fame and glory. But among them there are a few who have come to observe and to understand all that passes here.

It is the same with life. Some are influenced by the love of wealth while others are blindly led on by the mad fever for power and domination, but the finest type of man gives himself up to discovering the meaning and purpose of life itself. He seeks to uncover the secrets of nature. This is the man I call a philosopher for although no man is completely wise in all respects, he can love wisdom as the key to nature's secrets.

Although many were aware of Pythagoras' aspirations nobody outside of the Brotherhood knew the details or extent of his success. Each member of the school was forced to swear an oath never to reveal to the outside world any of their mathematical discoveries. Even after Pythagoras' death a member of the Brotherhood was drowned for breaking his oath – he publicly announced the discovery of a new regular solid, the dodecahedron, constructed from twelve regular pentagons. The highly secretive nature of the Pythagorean Brotherhood is part of the reason that myths have developed surrounding the strange rituals which they might have practised, and similarly this is why there are so few reliable accounts of their mathematical achievements.

What is known for certain is that Pythagoras established an ethos which changed the course of mathematics. The Brotherhood was effectively a religious community and one of the idols they worshipped was Number. By understanding the relationships between numbers, they believed that they could uncover the spiritual secrets of the universe and bring themselves closer to the gods. In particular the Brotherhood focused its attention on the study of counting numbers (1, 2, 3, …) and fractions. Counting numbers are sometimes called *whole numbers*, and together with fractions (ratios between whole numbers) are technically referred to as *rational numbers*. Among the infinity of numbers, the Brotherhood looked for those with special significance, and some of the most special were the so-called 'perfect' numbers.

According to Pythagoras numerical perfection depended on a number's divisors (numbers which will divide perfectly into the original one). For instance, the divisors of 12 are 1, 2, 3, 4 and 6. When the sum of a number's divisors is greater than the number itself, it is called an 'excessive' number. Therefore 12 is an excessive number because its divisors add up to 16. On the other hand, when the sum of a number's divisors is less than the number itself, it is called 'defective'. So 10 is a defective number because its divisors (1, 2 and 5) add up to only 8.

The most significant and rarest numbers are those whose divisors add up exactly to the number itself and these are the *perfect numbers*. The number 6 has the divisors 1, 2 and 3, and consequently it is a perfect number because $1 + 2 + 3 = 6$. The next perfect number is 28, because $1 + 2 + 4 + 7 + 14 = 28$.

As well as having mathematical significance for the Brotherhood, the perfection of 6 and 28 was acknowledged by other cultures who observed that the moon orbits the earth every 28 days and who declared that God created the world in 6 days. In *The City of God,* St Augustine argues that although God could have created the world in an instant he decided to take six days in order to reflect the universe's perfection. St Augustine observed that 6 was not perfect because God chose it, but rather that the perfection was inherent in the nature of the number: '6 is a number perfect in itself, and not because God created all things in six days; rather the inverse is true; God created all things in six days because this number is perfect. And it would remain perfect even if the work of the six days did not exist.'

As the counting numbers get bigger the perfect numbers become harder to find. The third perfect number is 496, the fourth is 8,128, the fifth is 33,550,336 and the sixth is 8,589,869,056. As well as being the sum of their divisors, Pythagoras noted that all perfect numbers exhibit several other elegant properties. For example, perfect numbers are always the sum of a series of consecutive counting numbers. So we have

$$6 = 1 + 2 + 3,$$
$$28 = 1 + 2 + 3 + 4 + 5 + 6 + 7,$$
$$496 = 1 + 2 + 3 + 4 + 5 + 6 + 7 + 8 + 9 + \cdots + 30 + 31,$$
$$8,128 = 1 + 2 + 3 + 4 + 5 + 6 + 7 + 8 + 9 + \cdots + 126 + 127.$$

Pythagoras was entertained by perfect numbers but he was not satisfied with merely collecting these special numbers; instead he desired to discover their deeper significance. One of his insights was that perfection was closely linked to 'twoness'. The numbers 4 (2×2), 8 ($2 \times 2 \times 2$), 16 ($2 \times 2 \times 2 \times 2$), etc., are known as powers of 2, and can be written as 2^n, where the n represents the number of 2's multiplied together. All these powers of 2 only just fail to be perfect, because the sum of their divisors always adds up to one less than the number itself. This makes them only slightly defective:

$$2^2 = 2 \times 2 \qquad\qquad\quad = \mathbf{4}, \quad \text{Divisors } 1, 2 \qquad\qquad\quad \text{Sum} = \mathbf{3},$$
$$2^3 = 2 \times 2 \times 2 \qquad\quad = \mathbf{8}, \quad \text{Divisors } 1, 2, 4 \qquad\qquad \text{Sum} = \mathbf{7},$$
$$2^4 = 2 \times 2 \times 2 \times 2 \quad = \mathbf{16}, \text{Divisors } 1, 2, 4, 8 \qquad\quad \text{Sum} = \mathbf{15},$$
$$2^5 = 2 \times 2 \times 2 \times 2 \times 2 = \mathbf{32}, \text{Divisors } 1, 2, 4, 8, 16 \ \text{Sum} = \mathbf{31}.$$

Two centuries later Euclid would refine Pythagoras' link between twoness and perfection. Euclid discovered that perfect numbers are always the multiple of two numbers, one of which is a power of 2 and the other being the next power of 2 minus 1. That is to say,

$$6 = 2^1 \times (2^2 - 1),$$
$$28 = 2^2 \times (2^3 - 1),$$
$$496 = 2^4 \times (2^5 - 1),$$
$$8,128 = 2^6 \times (2^7 - 1)$$

Today's computers have continued the search for perfect numbers and find such enormously large examples as $2^{216,090} \times (2^{216,091}-1)$, a number with over 130,000 digits, which obeys Euclid's rule.

Pythagoras was fascinated by the rich patterns and properties possessed by perfect numbers and respected their subtlety and cunning. At first sight perfection is a relatively simple concept to grasp and yet the ancient Greeks were unable to fathom some of the fundamental points of the subject. For example, although there are plenty of numbers whose divisors add up to one less than the number itself, that is to say only slightly defective, there appear to be no numbers which are slightly excessive. The Greeks were unable to find any numbers whose divisors added up to one more than the number itself, but they could not explain why this was the case. Frustratingly, although they failed to discover slightly excessive numbers, they could not prove that no such numbers existed. Understanding the apparent lack of slightly excessive numbers was of no practical value whatsoever; nonetheless it was a problem which might illuminate the nature of numbers and therefore it was worthy of study. Such riddles intrigued the Pythagorean Brotherhood, and two and a half thousand years later, mathematicians are still unable to prove that no slightly excessive numbers exist.

Everything is Number

In addition to studying the relationships within numbers Pythagoras was also intrigued by the link between numbers and nature. He realised that natural phenomena are governed by laws, and that these laws could be described by mathematical equations. One of the first links he discovered was the fundamental relationship between the harmony of music and the harmony of numbers.

The most important instrument in early Hellenic music was the tetrachord or four-stringed lyre. Prior to Pythagoras, musicians appreciated that particular notes when sounded together created a pleasant effect, and tuned their lyres so that plucking two strings would generate such a harmony. However, the early musicians had no understanding of why particular notes were harmonious and had no objective system for tuning their instruments. Instead they tuned their lyres purely by ear until a state of harmony was established – a process which Plato called torturing the tuning pegs.

Iamblichus, the fourth-century scholar who wrote nine books about the Pythagorean sect, describes how Pythagoras came to discover the underlying principles of musical harmony:

Once he was engrossed in the thought of whether he could devise a mechanical aid for the sense of hearing which would prove both certain and ingenious. Such an aid would be similar to the compasses, rules and optical instruments designed for the sense of sight. Likewise the sense of touch had scales and the concepts of weights and measures. By some divine stroke of luck he happened to walk past the forge of a blacksmith and listened to the hammers pounding iron and producing a variegated harmony of reverberations between them, except for one combination of sounds.

According to Iamblichus, Pythagoras immediately ran into the forge to investigate the harmony of the hammers. He noticed that most of the hammers could be struck simultaneously to generate a harmonious sound, whereas any combination containing one particular hammer always generated an unpleasant noise. He analysed the hammers and realised that those which were harmonious with each other had a simple mathematical relationship – their masses were simple ratios or fractions of each other. That is to say that hammers half, two-thirds or three-quarters the weight of a particular hammer would all generate harmonious sounds. On the other hand, the hammer which was generating disharmony when struck along with any of the other hammers had a weight which bore no simple relationship to the other weights.

Pythagoras had discovered that simple numerical ratios were responsible for harmony in music. Scientists have cast some doubt on Iamblichus' account of this story, but what is more certain is how Pythagoras applied his new theory of musical ratios to the lyre by examining the properties of a single string. Simply plucking the string generates a standard note or tone which is produced by the entire length of the vibrating string. By fixing the string at particular points along its length, it is possible to generate other vibrations

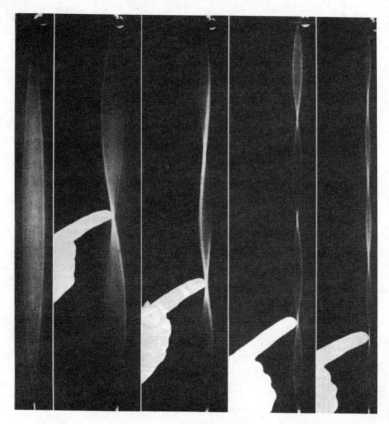

Figure 1. A freely vibrating open string generates a ground note. By creating a node exactly half-way along the string, the note generated is one octave higher and in harmony with the original note. Other harmonic notes can be generated by moving the node to other positions which are simple fractions (e.g. a third, a quarter, a fifth) of the distance along the string.

and tones, as illustrated in Figure 1. Crucially, harmonious tones only occur at very specific points. For example, by fixing the string at a point exactly half-way along it, plucking generates a tone which is one octave higher and in harmony with the original tone. Similarly, by fixing the string at points which are exactly a third, a quarter or a fifth of the way along it, other harmonious notes are produced. However, by fixing the string at a point which is not a simple fraction along the length of the whole string, a tone is generated which is not in harmony with the other tones.

Pythagoras had uncovered for the first time the mathematical rule which governs a physical phenomenon and demonstrated that there was a fundamental relationship between mathematics and science. Ever since this discovery scientists have searched for the mathematical rules which appear to govern every single physical process and have found that numbers crop up in all manner of natural phenomena. For example, one particular number appears to guide the lengths of meandering rivers. Professor Hans-Henrik Stølum, an earth scientist at Cambridge University, has calculated the ratio between the actual length of rivers from source to mouth and their direct length as the crow flies. Although the ratio varies from river to river, the average value is slightly greater than 3, that is to say that the actual length is roughly three times greater than the direct distance. In fact the ratio is approximately 3.14, which is close to the value of the number π, the ratio between the circumference of a circle and its diameter.

The number π was originally derived from the geometry of circles and yet it reappears over and over again in a variety of scientific circumstances. In the case of the river ratio, the appearance of π is the result of a battle between order and chaos. Einstein was the first to suggest that rivers have a tendency towards an ever more loopy path because the slightest curve will lead to faster currents on the outer side, which will in turn result in more erosion and a sharper bend. The sharper the bend, the faster the currents on the outer edge, the more the erosion, the more the river will twist, and so on. However, there is a natural process which will curtail the chaos: increasing loopiness will result in rivers doubling back on themselves and effectively short-circuiting. The river will become straighter and the loop will be left to one side forming an ox-bow lake. The balance between these two opposing factors leads to an average ratio of π between the actual length and the direct distance between source and mouth. The ratio of π is most commonly found for rivers flowing across very gently sloping plains, such as those found in Brazil or the Siberian tundra.

Pythagoras realised that numbers were hidden in everything, from the harmonies of music to the orbits of planets, and this led him to proclaim that 'Everything is Number'. By exploring the meaning of mathematics, Pythagoras was developing the language which would enable him and others to describe the nature of the universe. Henceforth each breakthrough in mathematics would give scientists the vocabulary they needed to better explain the phenomena around them. In fact developments in mathematics would inspire revolutions in science.

As well as discovering the law of gravity, Isaac Newton was a powerful mathematician. His greatest contribution to mathematics was his development of calculus, and in later years physicists would use the language of calculus to better describe the laws of gravity and to solve gravitational problems. Newton's classical theory of gravity survived intact for centuries until it was superseded by Albert Einstein's general theory of relativity, which developed a more detailed and alternative explanation of gravity. Einstein's own ideas were only possible because of new mathematical concepts which provided him with a more sophisticated

language for his more complex scientific ideas. Today the interpretation of gravity is once again being influenced by breakthroughs in mathematics. The very latest quantum theories of
gravity are tied to the development of mathematical strings, a
theory in which the geometrical and topological properties of tubes
seem to best explain the forces of nature.

Of all the links between numbers and nature studied by the
Brotherhood, the most important was the relationship which bears
their founder's name. Pythagoras' theorem provides us with an
equation which is true of all right-angled triangles and which
therefore also defines the right angle itself. In turn, the right angle
defines the perpendicular, i.e. the relation of the vertical to the
horizontal, and ultimately the relation between the three dimensions of our familiar universe. Mathematics, via the right angle,
defines the very structure of the space in which we live.

It is a profound realisation and yet the mathematics required to
grasp Pythagoras's theorem is relatively simple. To understand it,

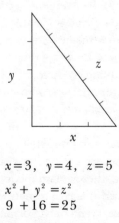

$$x = 3, \quad y = 4, \quad z = 5$$
$$x^2 + y^2 = z^2$$
$$9 + 16 = 25$$

Figure 2. All right-angled triangles obey Pythagoras' theorem.

simply begin by measuring the length of the two short sides of a right-angled triangle (x and y), and then square each one (x^2, y^2). Then add the two squared numbers ($x^2 + y^2$) to give you a final number. If you work out this number for the triangle shown in Figure 2, then the answer is 25.

You can now measure the longest side z, the so-called hypotenuse, and square this length. The remarkable result is that this number z^2 is identical to the one you just calculated, i.e. $5^2 =$ 25. That is to say,

> In a right-angled triangle the square on the hypotenuse is
> equal to the sum of the squares on the other two sides.

Or in other words (or rather symbols):

$$x^2 + y^2 = z^2.$$

This is clearly true for the triangle in Figure 2, but what is remarkable is that Pythagoras' theorem is true for every right-angled triangle you can possibly imagine. It is a universal law of mathematics, and you can rely on it whenever you come across any triangle with a right angle. Conversely if you have a triangle which obeys Pythagoras' theorem, then you can be absolutely confident that it is a right-angled triangle.

At this point it is important to note that, although this theorem will forever be associated with Pythagoras, it was actually used by the Chinese and the Babylonians one thousand years before. However, these cultures did not know that the theorem was true for every right-angled triangle. It was certainly true for the triangles they tested, but they had no way of showing that it was true for all the right-angled triangles which they had not tested. The reason for Pythagoras' claim to the theorem is that it was he who first demonstrated its universal truth.

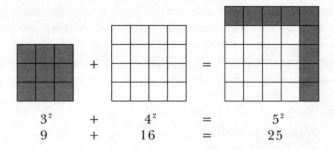

$$3^2 \quad + \quad 4^2 \quad = \quad 5^2$$
$$9 \quad + \quad 16 \quad = \quad 25$$

Figure 4. Finding whole number solutions to Pythagoras' equation can be thought of in terms of finding two squares which can be added together to form a third square. For example, a square made of 9 tiles can be added to a square of 16 tiles, and rearranged to form a third square made of 25 tiles.

Pythagorean triples are combinations of three whole numbers which perfectly fit Pythagoras' equation: $x^2 + y^2 = z^2$. For example, Pythagoras' equation holds true if $x = 3$, $y = 4$ and $z = 5$:

$$3^2 + 4^2 = 5^2, \qquad 9 + 16 = 25.$$

Another way to think of Pythagorean triples is in terms of re-arranging squares. If one has a 3×3 square made of 9 tiles, and a 4×4 square made of 16 tiles, then all the tiles can be rearranged to form a 5×5 square made of 25 tiles, as shown in Figure 4.

The Pythagoreans wanted to find other Pythagorean triples, other squares which could be added to form a third, larger square. Another Pythagorean triple is $x = 5$, $y = 12$ and $z = 13$:

$$5^2 + 12^2 = 13^2, \qquad 25 + 144 = 169.$$

A larger Pythagorean triple is $x = 99$, $y = 4,900$ and $z = 4,901$.

Pythagorean triples become rarer as the numbers increase, and finding them becomes harder and harder. To discover as many triples as possible the Pythgoreans invented a methodical way of finding them, and in so doing they also demonstrated that there are an infinite number of Pythagorean triples.

From Pythagoras' Theorem to Fermat's Last Theorem

Pythagoras' theorem and its infinity of triples was discussed in E.T. Bell's *The Last Problem*, the library book which caught the attention of the young Andrew Wiles. Although the Brotherhood had achieved an almost complete understanding of Pythagorean triples, Wiles soon discovered that this apparently innocent equation, $x^2 + y^2 = z^2$, has a darker side – Bell's book described the existence of a mathematical monster.

In Pythagoras' equation the three numbers, x, y and z, are all squared (i.e. $x^2 = x \times x$):

$$x^2 + y^2 = z^2.$$

However, the book described a sister equation in which x, y and z are all cubed (i.e. $x^3 = x \times x \times x$). The so-called power of x in this equation is no longer 2, but rather 3:

$$x^3 + y^3 = z^3.$$

Finding whole number solutions, i.e. Pythagorean triples, to the original equation was relatively easy, but changing the power from '2' to '3' (the square to a cube) and finding whole number solutions to the sister equation appears to be impossible. Generations of mathematicians scribbling on notepads have failed to find numbers which fit the equation perfectly.

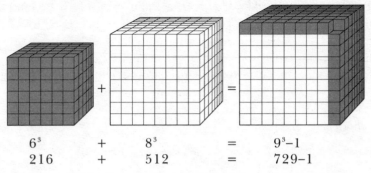

$$6^3 \quad\quad + \quad\quad 8^3 \quad\quad = \quad\quad 9^3{-}1$$
$$216 \quad\quad + \quad\quad 512 \quad\quad = \quad\quad 729{-}1$$

Figure 5. Is it possible to add the building blocks from one cube to another cube, to form a third, larger, cube? In this case a 6 × 6 × 6 cube added to an 8 × 8 × 8 cube does not have quite enough building blocks to form a 9 × 9 × 9 cube. There are 216 (6^3) building blocks in the first cube, and 512 (8^3) in the second. The total is 728 building blocks, which is 1 short of 9^3.

With the original 'squared' equation, the challenge was to re-arrange the tiles in two squares to form a third, larger square. The 'cubed' version of the challenge is to rearrange two cubes made of building blocks, to form a third, larger cube. Apparently, no matter what cubes are chosen to begin with, when they are combined the result is either a complete cube with some extra blocks left over, or an incomplete cube. The nearest that anyone has come to a per-fect rearrangement is one in which there is one building block too many or too few. For example, if we begin with the cubes 6^3 (x^3) and 8^3 (y^3) and rearrange the building blocks, then we are only one short of making a complete 9 × 9 × 9 cube, as shown in Figure 5.

Finding three numbers which fit the cubed equation perfectly seems to be impossible. That is to say, there appear to be no whole number solutions to the equation

$$x^3 + y^3 = z^3.$$

Furthermore, if the power is changed from 3 (cubed) to any higher number n (i.e. 4, 5, 6, …), then finding a solution still seems to be impossible. There appear to be no whole number solutions to the more general equation

$$x^n + y^n = z^n \quad \text{for } n \text{ greater than 2.}$$

By merely changing the 2 in Pythagoras' equation to any higher number, finding whole number solutions turns from being relatively simple to being mind-bogglingly difficult. In fact, the great seventeenth-century Frenchman Pierre de Fermat made the astonishing claim that the reason why nobody could find any solutions was that no solutions existed.

Fermat was one of the most brilliant and intriguing mathematicians in history. He could not have checked the infinity of numbers, but he was absolutely sure that no combination existed which would fit the equation perfectly because his claim was based on proof. Like Pythagoras, who did not have to check every triangle to demonstrate the validity of his theorem, Fermat did not have to check every number to show the validity of his theorem. Fermat's Last Theorem, as it is known, stated that

$$x^n + y^n = z^n$$

has no whole number solutions for n greater than 2.

As Wiles read each chapter of Bell's book, he learnt how Fermat had become fascinated by Pythagoras' work and had eventually come to study the perverted form of Pythagoras' equation. He then read how Fermat had claimed that even if all the mathematicians in the world spent eternity looking for a solution to the equation they would fail to find one. He must have eagerly turned the pages, relishing the thought of examining the proof of Fermat's Last Theorem. However, the proof was not there. It was not anywhere. Bell ended the book by stating that the proof had been lost long ago. There was no hint of what it might have been, no clues as to the proof's construction or derivation. Wiles found himself puzzled, infuriated and intrigued. He was in good company.

For over 300 years many of the greatest mathematicians had tried to rediscover Fermat's lost proof and failed. As each generation failed, the next became even more frustrated and determined. In 1742, almost a century after Fermat's death, the Swiss mathematician Leonhard Euler asked his friend Clêrot to search Fermat's house in case some vital scrap of paper still remained. No clues were ever found as to what Fermat's proof might have been. In Chapter 2 we shall find out more about the mysterious Pierre de Fermat and how his theorem came to be lost, but for the time being it is enough to know that Fermat's Last Theorem, a problem that had captivated mathematicians for centuries, had captured the imagination of the young Andrew Wiles.

Sat in Milton Road Library was a ten-year-old boy staring at the most infamous problem in mathematics. Usually half the difficulty in a mathematics problem is understanding the question, but in this case it was simple – prove that $x^n + y^n = z^n$ has no whole number solutions for n greater than 2. Andrew was not daunted by the knowledge that the most brilliant minds on the planet had failed to rediscover the proof. He immediately set to work using all his textbook techniques to try and recreate the proof. Perhaps he could find something that everyone else, except Fermat, had overlooked. He dreamed he could shock the world.

Thirty years later Andrew Wiles was ready. Standing in the

On 23 June 1993 Wiles gave a lecture at the Isaac Newton Institute in Cambridge. This was the moment immediately after he announced his proof of Fermat's Last Theorem. He, along with everyone else in the room, had no idea of the nightmare ahead.

auditorium of the Isaac Newton Institute, he scribbled on the board and then, struggling to contain his glee, stared at his audience. The lecture was reaching its climax and the audience knew it. One or two of them had smuggled cameras into the lecture room and flashes peppered his concluding remarks.

With the chalk in his hand he turned to the board for the last time. The final few lines of logic completed the proof. For the first time in over three centuries Fermat's challenge had been met. A few more cameras flashed to capture the historic moment. Wiles wrote up the statement of Fermat's Last Theorem, turned towards the audience, and said modestly: 'I think I'll stop here.'

Two hundred mathematicians clapped and cheered in celebration. Even those who had anticipated the result grinned in disbelief. After three decades Andrew Wiles believed he had achieved his dream, and after seven years of isolation he could reveal his secret calculation. However, while euphoria filled the Newton Institute tragedy was about to strike. As Wiles was enjoying the moment, he, along with everyone else in the room, was oblivious of the horrors to come.

Pierre de Fermat

2

The Riddler

'Do you know,' the Devil confided, 'not even the best mathematicians on other planets – all far ahead of yours – have
solved it? Why, there's a chap on Saturn – he looks something like a mushroom on stilts – who solves partial
differential equations mentally; and even he's given up.'

Arthur Poges, 'The Devil and Simon Flagg'

Pierre de Fermat was born on 20 August 1601 in the town of
Beaumont-de-Lomagne in south-west France. Fermat's father,
Dominique Fermat, was a wealthy leather merchant, and so Pierre
was fortunate enough to enjoy a privileged education at the
Franciscan monastery of Grandselve, followed by a stint at the
University of Toulouse. There is no record of the young Fermat
showing any particular brilliance in mathematics.

Pressure from his family steered Fermat towards a career in the
civil service, and in 1631 he was appointed *conseiller au Parlement de
Toulouse*, a councillor at the Chamber of Petitions. If locals wanted
to petition the King on any matter they first had to convince
Fermat or one of his associates of the importance of their request.
The councillors provided the vital link between the province and
Paris. As well as liaising between the locals and the monarch, the
councillors made sure that royal decrees emanating from the capital were implemented back in the regions. Fermat was an efficient

civil servant, who by all accounts carried out his duties in a considerate and merciful manner.

Fermat's additional duties included service in the judiciary and he was senior enough to deal with the most severe cases. An account of his work is given by the English mathematician, Sir Kenelm Digby. Digby had requested to see Fermat, but in a letter to a mutual colleague, John Wallis, he reveals that the Frenchman had been occupied with pressing judicial matters, thus excluding the possibility of a meeting:

It is true that I had exactly hit the date of the displacement of the judges of Castres to Toulouse, where he [Fermat] is the Supreme Judge to the Sovereign Court of Parliament; and since then he has been occupied with capital cases of great importance, in which he has finished by imposing a sentence that has made a great stir; it concerned the condemnation of a priest, who had abused his functions, to be burned at the stake. This affair has just finished and the execution has followed.

Fermat corresponded regularly with Digby and Wallis. Later we will see that the letters were often less than friendly, but they provide vital insights into Fermat's daily life, including his academic work.

Fermat rose rapidly within the ranks of the civil service and became a member of the social élite, entitling him to use *de* as part of his name. His promotion was not necessarily the result of ambition, but rather a matter of health. The plague was raging throughout Europe and those who survived were elevated to fill the places of those who died. Even Fermat suffered a serious bout of plague in 1652, and was so ill that his friend Bernard Medon announced his death to several colleagues. Soon after he corrected himself in a report to the Dutchman Nicholas Heinsius:

I informed you earlier of the death of Fermat. He is still alive, and we no longer fear for his health, even though we had counted him among the dead a short time ago. The plague no longer rages among us.

In addition to the health risks of seventeenth-century France, Fermat had to survive the political dangers. His appointment to the Parliament of Toulouse came just three years after Cardinal Richelieu was promoted to first minister of France. This was an era of plotting and intrigue, and everyone involved in the running of the state, even at local government level, had to take care not to become embroiled in the machinations of the Cardinal. Fermat adopted the strategy of performing duties efficiently without drawing attention to himself. He had no great political ambition, and did his best to avoid the rough and tumble of parliament. Instead he devoted all his spare energy to mathematics and, when not sentencing priests to be burnt at the stake, Fermat dedicated himself to his hobby. Fermat was a true amateur academic, a man whom E.T. Bell called the 'Prince of Amateurs'. But so great were his talents that when Julian Coolidge wrote *Mathematics of Great Amateurs*, he excluded Fermat on the grounds that he was 'so really great that he should count as a professional'.

At the start of the seventeenth century, mathematics was still recovering from the Dark Ages and was not a highly regarded subject. Similarly mathematicians were not treated with great respect and most of them had to fund their own studies. For example, Galileo was unable to study mathematics at the University of Pisa and was forced to seek private tuition. Indeed, the only institute in Europe to actively encourage mathematicians was Oxford University which had established the Savilian Chair of Geometry in 1619. It is true to say that most seventeenth-century mathematicians were amateurs, but Fermat was an extreme case. Living far from Paris he was isolated from the small community of mathematicians that did exist, which included such figures as Pascal, Gassendi, Roberval, Beaugrand and most notably Father Marin Mersenne.

Father Mersenne made only minor contributions to number theory and yet he played a role in seventeenth-century mathematics which was arguably more important than any of his more esteemed colleagues. After joining the order of Minims in 1611, Mersenne studied mathematics and then taught the subject to other monks and to nuns at the Minim convent at Nevers. Eight years later he moved to Paris to join the Minims de l'Annociade, close to the Place Royale, a natural gathering place for intellectuals. Inevitably Mersenne met the other mathematicians of Paris, but he was saddened by their reluctance to talk to him or to each other.

The secretive nature of the Parisian mathematicians was a tradition which had been passed down from the cossists of the sixteenth century. The cossists were experts in calculations of all kinds and were employed by merchants and businessmen to solve complex accounting problems. Their name derives from the Italian word *cosa*, meaning 'thing', because they used symbols to represent an unknown quantity, similar to the way mathematicians use x today. All professional problem-solvers of this era invented their own clever methods for performing calculations and would do their utmost to keep these methods secret in order to maintain their reputation as the only person capable of solving a particular problem. On one exceptional occasion Niccolò Tartaglia, who had found a method for quickly solving cubic equations, revealed his discovery to Girolamo Cardano and swore him to absolute secrecy. Ten years later Cardano broke his promise and published Tartaglia's method in his *Ars Magna*, an act which Tartaglia would never forgive. He broke off all relations with Cardano and a bitter public dispute ensued, which only served to further encourage other mathematicians to guard their secrets. The secretive nature of mathematicians continued right up until the end of the nineteenth century, and as we shall see later there are even examples of secret geniuses working in the twentieth century.

When Father Mersenne arrived in Paris he was determined to fight against the ethos of secrecy and tried to encourage mathematicians to exchange their ideas and build upon each other's work. The monk arranged regular meetings and his group later formed the core of the French Academy. When anyone refused to attend, Mersenne would pass on to the group whatever he could by revealing letters and papers – even if they had been sent to him in confidence. It was not ethical behaviour for a man of the cloth, but he justified it on the grounds that the exchange of information would benefit mathematics and mankind. These acts of indiscretion naturally caused bitter arguments between the well-meaning monk and the taciturn prima donnas, and eventually destroyed Mersenne's relationship with Descartes which had lasted since the two men had studied together at the Jesuit College of La Flèche. Mersenne had revealed philosophical writings by Descartes which were liable to offend the Church, but to his credit he did defend Descartes against theological attacks, as in fact he had done earlier in the case of Galileo. In an era dominated by religion and magic Mersenne stood up for rational thought.

Mersenne travelled throughout France and further afield, spreading news of the latest discoveries. In his travels he would make a point of meeting up with Pierre de Fermat and, indeed, seems to have been Fermat's only regular contact with other mathematicians. Mersenne's influence on this Prince of Amateurs must have been second only to the *Arithmetica*, a mathematical treatise handed down from the ancient Greeks which was Fermat's constant companion. Even when he was unable to travel Mersenne would maintain his relationship with Fermat and others by writing prolifically. After Mersenne's death his room was found stacked with letters written by seventy-eight different correspondents.

Despite the encouragement of Father Mersenne, Fermat steadfastly refused to reveal his proofs. Publication and recognition meant nothing to him and he was satisfied with the simple pleasure of being able to create new theorems undisturbed. However, the shy and retiring genius did have a mischievous streak, which, when combined with his secrecy, meant that when he did sometimes communicate with other mathematicians it was only to tease them. He would write letters stating his most recent theorem without providing the accompanying proof. Then he would challenge his contemporaries to find the proof. The fact that he would never reveal his own proofs caused a great deal of frustration. René Descartes called Fermat a 'braggart' and the Englishman John Wallis referred to him as 'That damned Frenchman'. Unfortunately for the English, Fermat took particular pleasure in toying with his cousins across the Channel.

As well as having the satisfaction of annoying his colleagues, Fermat's habit of stating a problem but hiding its solution did have more practical motivations. First, it meant that he did not have to waste time fully fleshing out his methods; instead he could rapidly proceed to his next conquest. Furthermore, he did not have to suffer jealous nit-picking. Once published, proofs would be examined and argued over by everyone and anyone who knew anything about the subject. When Blaise Pascal pressed him to publish some of his work, the recluse replied: 'Whatever of my work is judged worthy of publication, I do not want my name to appear there.'

Fermat was the secretive genius who sacrificed fame in order not to be distracted by petty questions from his critics.

This exchange of letters with Pascal, the only occasion when Fermat discussed ideas with anyone but Mersenne, concerned the creation of an entirely new branch of mathematics – probability theory. The mathematical hermit was introduced to the subject by Pascal, and so, despite his desire for isolation, he felt obliged to maintain a dialogue. Together Fermat and Pascal would discover the first proofs and cast-iron certainties in probability theory, a subject which is inherently uncertain. Pascal's interest in the subject had been sparked by a professional Parisian gambler, Antoine Gombaud, the Chevalier de Méré, who had posed a problem which concerned a game of chance called *points*. The game involves winning points on the roll of a dice, and whichever player is the first to earn a certain number of points is the winner and takes the prize money.

Gombaud had been involved in a game of points with a fellow-gambler when they were forced to abandon the game half-way through, owing to a pressing engagement. The problem then arose as to what to do with the prize money. The simple solution would have been to have given all the money to the competitor with the most points, but Gombaud asked Pascal if there was a fairer way to divide the money. Pascal was asked to calculate the probability of each player winning had the game continued and assuming that both players would have had an equal chance of winning subsequent points. The prize money could then be split according to these calculated probabilities.

Prior to the seventeenth century the laws of probability were defined by the intuition and experience of gamblers, but Pascal entered into an exchange of letters with Fermat with the aim of discovering the mathematical rules which more accurately describe the laws of chance. Three centuries later Bertrand Russell would comment on this apparent oxymoron: 'How dare we speak of the laws of chance? Is not chance the antithesis of all law?'

The Frenchmen analysed Gombaud's question and soon realised that it was a relatively trivial problem which could be solved by rigorously defining all the potential outcomes of the game and assigning an individual probability to each one. Both Pascal and Fermat were capable of independently solving Gombaud's problem, but their collaboration speeded up the discovery of a solution and led them to a deeper exploration of other more subtle and sophisticated questions related to probability.

Probability problems are sometimes controversial because the mathematical answer, the true answer, is often contrary to what intuition might suggest. This failure of intuition is perhaps surprising because 'survival of the fittest' ought to provide a strong evolutionary pressure in favour of a brain naturally capable of analysing questions of probability. You can imagine our ancestors stalking a young deer, and weighing up whether or not to attack. What is the risk that a stag is nearby ready to defend its offspring and injure its assailant? On the other hand what is the chance that a better opportunity for a meal might arise if this one is judged too risky? A talent for analysing probability should be part of our genetic make-up and yet often our intuition misleads us.

One of the most counterintuitive probability problems concerns the likelihood of sharing birthdays. Imagine a football pitch with 23 people on it, the players and the referee. What is the probability that any two of those 23 people share the same birthday? With 23 people and 365 birthdays to chose from, it would seem highly unlikely that anybody would share the same birthday. If asked to put a figure on it most people would guess a probability of perhaps 10% at most. In fact, the actual answer is just over 50% – that is to say, on the balance of probability, it is more likely than not that two people on the pitch will share the same birthday.

The reason for this high probability is that what matters more than the number of people is the number of ways people can be paired. When we look for a shared birthday, we need to look at pairs of people not individuals. Whereas there are only 23 people on the pitch, there are 253 pairs of people. For example, the first person can be paired with any of the other 22 people giving 22 pairings to start with. Then, the second person can be paired with any of the remaining 21 people (we have already counted the second person paired with the first person so the number of possible pairings is reduced by one), giving an additional 21 pairings. Then, the third person can be paired with any of the remaining 20 people, giving an additional 20 pairings, and so on until we reach a total of 253 pairs.

The fact that the probability of a shared birthday within a group of 23 people is more than 50% seems intuitively wrong, and yet it is mathematically undeniable. Strange probabilities such as this are exactly what bookmakers and gamblers rely on in order to exploit the unwary. The next time you are at a party with more than 23 people you might want to make a wager that two people in the room will share a birthday. Please note that with a group of 23 people the probability is only slightly more than 50%, but the probability rapidly rises as the group increases in size. Hence, with a party of 30 people it is certainly worth betting that two of them will share the same birthday.

Fermat and Pascal founded the essential rules which govern all games of chance and which can be used by gamblers to define perfect playing and betting strategies. Furthermore, these laws of probability have found applications in a whole series of situations, ranging from speculating on the stock market to estimating the probability of a nuclear accident. Pascal was even convinced that he could use his theories to justify a belief in God. He stated that 'the excitement that a gambler feels when making a bet is equal to the amount he might win multiplied by the probability of winning it'. He then argued that the possible prize of eternal happiness has an infinite value and that the probability of entering heaven by leading a virtuous life, no matter how small, is certainly finite. Therefore, according to Pascal's definition, religion was a game of infinite excitement and one worth playing, because multiplying an infinite prize by a finite probability results in infinity.

As well as sharing the parentage of probability theory, Fermat was deeply involved in the founding of another area of mathematics, calculus. Calculus is the ability to calculate the rate of change, known as the derivative, of one quantity with respect to another. For example, the rate of change of distance with respect to time is better known simply as velocity. For mathematicians the quantities tend to be abstract and intangible but the consequences of Fermat's work were to revolutionise science. Fermat's mathematics enabled scientists to better understand the concept of velocity and its relation to other fundamental quantities such as acceleration – the rate of change of velocity with respect to time.

Economics is a subject heavily influenced by calculus. Inflation is the rate of change of price, known as the derivative of price, and furthermore economists are often interested in the rate of change of inflation, known as the second derivative of price. These terms are frequently used by politicians and the mathematician Hugo Rossi once observed the following: 'In the fall of 1972 President Nixon announced that the rate of increase of inflation was decreasing. This was the first time a sitting president used a third derivative to advance his case for re-election.'

For centuries Isaac Newton was thought to have discovered calculus independently and without any knowledge of Fermat's work, but in 1934 Louis Trenchard Moore discovered a note which set the record straight and gave Fermat the credit he deserves. Newton wrote that he developed his calculus based on 'Monsieur Fermat's method of drawing tangents'. Ever since the seventeenth century calculus has been used to describe Newton's law of gravity and his laws of mechanics, which depend on distance, velocity and acceleration.

The discovery of calculus and probability theory would have been more than enough to earn Fermat a place in the mathematicians' hall of fame, but his greatest achievement was in yet another branch of mathematics. While calculus has since been used to send rockets to the moon, and while probability theory has been used for risk assessment by insurance companies, Fermat's greatest love was for a subject which is largely useless – the theory of numbers. Fermat was driven by an obsession to understand the properties of and the relationships between numbers. This is the purest and most ancient form of mathematics and Fermat was building on a body of knowledge that had been handed down to him from Pythagoras.

The Evolution of Number Theory

After Pythagoras' death the concept of mathematical proof rapidly spread across the civilised world, and two centuries after his School was burnt to the ground the hub of mathematical study had moved from Croton to the city of Alexandria. In 332 BC, having conquered Greece, Asia Minor and Egypt, Alexander the Great decided that he would build a capital city that would be the most magnificent in the world. Alexandria was indeed a spectacular metropolis but not immediately a centre of learning. It was only when Alexander died and his half-brother Ptolemy I ascended the throne of Egypt that Alexandria became home to the world's first-ever university. Mathematicians and other intellectuals flocked to Ptolemy's city of culture, and although they were certainly drawn by the reputation of the university, the main attraction was the Alexandrian Library.

The Library was the idea of Demetrius Phalaerus, an unpopular orator who had been forced to flee Athens, and who eventually found sanctuary in Alexandria. He persuaded Ptolemy to gather together all the great books, assuring him that the great minds would follow. Once the tomes of Egypt and Greece had been installed, agents scoured Europe and Asia Minor in search of further volumes of knowledge. Even tourists to Alexandria could not escape the voracious appetite of the Library. Upon entering the city, their books were confiscated and taken to the scribes. The books were copied so that while the original was donated to the Library, a duplicate could graciously be given to the original owner. This meticulous replication service for ancient travellers gives today's historians some hope that a copy of a great lost text will one day turn up in an attic somewhere in the world. In 1906 J.L. Heiberg discovered in Constantinople just such a manuscript, *The Method*, which contained some of Archimedes' original writings.

Ptolemy's dream of building a treasure house of knowledge lived on after his death, and by the time a few more Ptolemys had ascended the throne the Library contained over 600,000 books. Mathematicians could learn everything in the known world by studying at Alexandria, and there to teach them were the most famous academics. The first head of the mathematics department was none other than Euclid.

Euclid was born in about 330 BC. Like Pythagoras, Euclid believed in the search for mathematical truth for its own sake and did not look for applications in his work. One story tells of a student who questioned him about the use of the mathematics he was learning. Upon completing the lesson, Euclid turned to his slave and said, 'Give the boy a penny since he desires to profit from all that he learns.' The student was then expelled.

Euclid devoted much of his life to writing the *Elements*, the most successful textbook in history. Until this century it was also the second best-selling book in the world after the Bible. The *Elements* consists of thirteen books, some of which are devoted to Euclid's own work, and the remainder being a compilation of all the mathematical knowledge of the age, including two volumes devoted entirely to the works of the Pythagorean Brotherhood. In the centuries since Pythagoras, mathematicians had invented a variety of logical techniques which could be applied in different circumstances, and Euclid skilfully employed them all in the *Elements*. In particular Euclid exploited a logical weapon known as *reductio ad absurdum*, or proof by contradiction. The approach revolves around the perverse idea of trying to prove that a theorem is true by first assuming that the theorem is false. The mathematician then explores the logical consequences of the theorem being false. At some point along the chain of logic there is a contradiction (e.g. $2 + 2 = 5$). Mathematics abhors a contradiction and therefore the original theorem cannot be false, i.e. it must be true.

The English mathematician G.H. Hardy encapsulated the spirit of proof by contradiction in his book *A Mathematician's Apology*: 'Reductio ad absurdum, which Euclid loved so much, is one of a mathematician's finest weapons. It is a far finer gambit than any chess play: a chess player may offer the sacrifice of a pawn or even a piece, but a mathematician offers the game.'

One of Euclid's most famous proofs by contradiction established the existence of so-called *irrational numbers*. It is suspected that irrational numbers were originally discovered by the Pythagorean Brotherhood centuries earlier, but the concept was so abhorrent to Pythagoras that he denied their existence.

When Pythagoras claimed that the universe is governed by numbers he meant whole numbers and ratios of whole numbers (fractions) together known as rational numbers. An irrational number is a number that is neither a whole number nor a fraction, and this is what made it so horrific to Pythagoras. In fact, irrational numbers are so strange that they cannot be written down as decimals, even recurring decimals. A recurring decimal such as 0.111111… is in fact a fairly straightforward number, and is equivalent to the fraction $\frac{1}{9}$. The fact that the '1' repeats itself forever means that the decimal has a very simple and regular pattern. This regularity, despite the fact that it continues to infinity, means that the decimal can be rewritten as a fraction. However, if you attempt to express an irrational number as a decimal you end up with a number which continues forever with no regular or consistent pattern.

The concept of an irrational number was a tremendous breakthrough. Mathematicians were looking beyond the whole numbers and fractions around them, and discovering, or perhaps inventing, new ones. The nineteenth-century mathematician Leopold Kronecker said, 'God made the integers; all the rest is the work of man.'

The most famous irrational number is π. In schools it is sometimes approximated by $3\frac{1}{7}$ or 3.14; however, the true value of π is nearer 3.14159265358979323846, but even this is only an approximation. In fact, π can never be written down exactly because the decimal places go on forever without any pattern. A beautiful feature of this random pattern is that it can be computed using an equation which is supremely regular:

$$\pi = 4 \left(\frac{1}{1} - \frac{1}{3} + \frac{1}{5} - \frac{1}{7} + \frac{1}{9} - \frac{1}{11} + \frac{1}{13} - \frac{1}{15} + \cdots \right) .$$

By calculating the first few terms, you can obtain a very rough value for π, but by calculating more and more terms an increasingly accurate value is achieved. Although knowing π to 39 decimal places is sufficient to calculate the circumference of the universe accurate to the radius of a hydrogen atom, this has not prevented computer scientists from calculating π to as many decimal places as possible. The current record is held by Yasumasa Kanada of the University of Tokyo who calculated π to six billion decimal places in 1996. Recently rumours have suggested that the Russian Chudnovsky brothers in New York have calculated π to eight billion decimal places and that they are aiming to reach a trillion decimal places. However, even if Kanada or the Chudnovsky brothers carried on calculating until their computers sapped all the energy in the universe, they would still not have found the exact value of π. It is easy to appreciate why Pythagoras conspired to hide the existence of these mathematical beasts.

The value of π to over 1500 decimal places

3.14159265358979323846264338327950288419716939937510582
09749445923078164062862089986280348253421170679821480865
13282306647093844609550582231725359408128481117450284102
70193852110555964462294895493038196442881097566593344612
84756482337867831652712019091456485669234603486104543266
48213393607260249141273724587006606315588174881520920962
82925409171536436789259036001133053054882046652138414695
19415116094330572703657595919530921861173819326117931051
18548074462379962749567351885752724891227938183011194912
98336733624406566430860213949463952247371907021798609437
02770539217176293176752384674818467669405132000568127145
26356082778577134275778960917363717872146844090122495343
01465495853710507922796892589235420199561121290219608640
34418159813629774771309960518707211349999998372978049951
05973173281609631859502445945533469083026425223082533446
85035261931188171010003137838752886587533208381420617177
66914730359825349042875546873115956286388235378759375195
77818577805321712268066130019278766111959092164201989380
95257201065485863278865936153381827968230301952035301852
96899577362259941389124972177528347913151557485724245415
06959508295331168617278558890750983817546374649393192550
60400927701671139009848824012858361603563707660104710181
94295559619894676783744944825537977472684710404753446462
08046684259069491293313677028989152104752162056966024058
03815019351125338243003558764024749647326391419927260426
99227967823547816360093417216412199245863150302861829745
55706749838505494588586926995690927210797509302955321165
34498720275596023648066549119881834797753566369807426542
52786255181841757467289097777279380008164702001614524919
21732172147723501414419735

When Euclid dared to confront the issue of irrationality in the tenth volume of the *Elements* the goal was to prove that there could be a number which could never be written as a fraction. Instead of trying to prove that π is irrational, he examined the square root of two, $\sqrt{2}$ – the number which when multiplied by itself is equal to two. In order to prove that $\sqrt{2}$ could not be written as a fraction Euclid used *reductio ad absurdum* and began by assuming that it could be written as a fraction. He then demonstrated that this hypothetical fraction could be simplified. Simplification of a fraction means, for example, that the fraction $\frac{8}{12}$ can be simplified to $\frac{4}{6}$ by dividing top and bottom by 2. In turn $\frac{4}{6}$ can be simplified to $\frac{2}{3}$, which cannot be simplified any further and therefore the fraction is then said to be in its simplest form. However, Euclid showed that his hypothetical fraction, which was supposed to represent $\sqrt{2}$, could be simplified not just once, but over and over again an infinite number of times without ever reducing to its simplest form. This is absurd because all fractions must eventually have a simplest form, and therefore the hypothetical fraction cannot exist. Therefore $\sqrt{2}$ cannot be written as a fraction and is irrational. An outline of Euclid's proof is given in Appendix 2.

By using proof by contradiction Euclid was able to prove the existence of irrational numbers. For the first time numbers had taken on a new and more abstract quality. Until this point in history all numbers could be expressed as whole numbers or fractions, but Euclid's irrational numbers defied representation in the traditional manner. There is no other way to describe the number equal to the square root of two other than by expressing it as $\sqrt{2}$, because it cannot be written as a fraction and any attempt to write it as a decimal could only ever be an approximation, e.g. 1.414213562373...

For Pythagoras, the beauty of mathematics was the idea that

rational numbers (whole numbers and fractions) could explain all natural phenomena. This guiding philosophy blinded Pythagoras to the existence of irrational numbers and may even have led to the execution of one of his pupils. One story claims that a young student by the name of Hippasus was idly toying with the number $\sqrt{2}$, attempting to find the equivalent fraction. Eventually he came to realise that no such fraction existed, i.e. that $\sqrt{2}$ is an irrational number. Hippasus must have been overjoyed by his discovery, but his master was not. Pythagoras had defined the universe in terms of rational numbers, and the existence of irrational numbers brought his ideal into question. The consequence of Hippasus' insight should have been a period of discussion and contemplation during which Pythagoras ought to have come to terms with this new source of numbers. However, Pythagoras was unwilling to accept that he was wrong, but at the same time he was unable to destroy Hippasus' argument by the power of logic. To his eternal shame he sentenced Hippasus to death by drowning.

The father of logic and the mathematical method had resorted to force rather than admit he was wrong. Pythagoras' denial of irrational numbers is his most disgraceful act and perhaps the greatest tragedy of Greek mathematics. It was only after his death that irrationals could be safely resurrected.

Although Euclid clearly had an interest in the theory of numbers, it was not his greatest contribution to mathematics. Euclid's true passion was geometry, and of the thirteen volumes that make up the *Elements*, books I to VI concentrate on plane (two-dimensional) geometry and books XI to XIII deal with solid (three-dimensional) geometry. It is such a complete body of knowledge that the contents of the *Elements* would form the geometry syllabus in schools and universities for the next two thousand years.

The mathematician who compiled the equivalent text for

number theory was Diophantus of Alexandria, the last champion of the Greek mathematical tradition. Although Diophantus' achievements in number theory are well documented in his books, virtually nothing else is known about this formidable mathematician. His place of birth is unknown and his arrival in Alexandria could have been any time within a five-century window. In his writings Diophantus quotes Hypsicles and therefore he must have lived after 150 BC; on the other hand his own work is quoted by Theon of Alexandria and therefore he must have lived before AD 364. A date around AD 250 is generally accepted as being a sensible estimate. Appropriately for a problem-solver, the one detail of Diophantus' life that has survived is in the form of a riddle said to have been carved on his tomb:

God granted him to be a boy for the sixth part of his life, and adding a twelfth part to this, He clothed his cheeks with down; He lit him the light of wedlock after a seventh part, and five years after his marriage He granted him a son. Alas! late-born wretched child; after attaining the measure of half his father's full life, chill Fate took him. After consoling his grief by this science of numbers for four years he ended his life.

The challenge is to calculate Diophantus' life span. The answer can be found in Appendix 3.

This riddle is an example of the sort of problem that Diophantus relished. His speciality was to tackle questions which required whole number solutions, and today such questions are referred to as Diophantine problems. He spent his career in Alexandria collecting well-understood problems and inventing new ones, and then compiled them all into a major treatise entitled *Arithmetica*. Of the thirteen books which made up the *Arithmetica*, only six would survive the turmoils of the Dark Ages and go on to inspire the Renaissance mathematicians, including Pierre de Fermat. The

The frontispiece of Claude Gaspar Bachet's translation of Diophantus' *Arithmetica*, published in 1621. This book became Fermat's bible and inspired much of his work.

remaining seven books would be lost during a series of tragic events which would send mathematics back to the age of the Babylonians.

During the centuries between Euclid and Diophantus, Alexandria remained the intellectual capital of the civilised world, but throughout this period the city was continually under threat from foreign armies. The first major attack occurred in 47 BC, when Julius Caesar attempted to overthrow Cleopatra by setting fire to the Alexandrian fleet. The Library, which was located near the harbour, also caught alight, and hundreds of thousands of books were destroyed. Fortunately for mathematics Cleopatra appreciated the importance of knowledge and was determined to restore the Library to its former glory. Mark Antony realised that the way to an intellectual's heart is via her library, and so marched to the city of Pergamum. This city had already initiated a library which it hoped would provide it with the best collection in the world, but instead Mark Antony transplanted the entire stock to Egypt, restoring the supremacy of Alexandria.

For the next four centuries the Library continued to accumulate books until in AD 389 it received the first of two fatal blows, both the result of religious bigotry. The Christian Emperor Theodosius ordered Theophilus, Bishop of Alexandria, to destroy all pagan monuments. Unfortunately when Cleopatra rebuilt and restocked the Library, she decided to house it in the Temple of Serapis, and so the Library became caught up in the destruction of icons and altars. The 'pagan' scholars attempted to save six centuries-worth of knowledge, but before they could do anything they were butchered by the Christian mob. The descent into the Dark Ages had begun.

A few precious copies of the most vital books survived the Christian onslaught and scholars continued to visit Alexandria in search of knowledge. Then in 642 a Moslem attack succeeded where the Christians had failed. When asked what should be done with the Library, the victorious Caliph Omar commanded that those books that were contrary to the Koran should be destroyed, and furthermore those books that conformed to the Koran were superfluous and they too must be destroyed. The manuscripts were used to stoke the furnaces which heated the public baths and Greek mathematics went up in smoke. It is not surprising that most of Diophantus' work was destroyed; in fact it is a miracle that six volumes of the *Arithmetica* managed to survive the tragedy of Alexandria.

For the next thousand years mathematics in the West was in the doldrums, and only a handful of luminaries in India and Arabia kept the subject alive. They copied the formulae described in the surviving manuscripts of Greece and then began to reinvent for themselves many of the theorems that had been lost. They also added new elements to mathematics, including the number zero.

In modern mathematics zero performs two functions. First, it allows us to distinguish between numbers like 52 and 502. In a system where the position of a number denotes its value, a symbol is needed to confirm an empty position. For instance, 52 represents 5 times ten plus 2 times one, whereas 502 represents 5 times a hundred plus 0 times ten plus 2 times one, and the zero is crucial for removing any ambiguity. Even the Babylonians in the third millennium BC appreciated the use of zero to avoid confusion, and the Greeks adopted their idea, using a circular symbol similar to the one we use today. However, zero has a more subtle and deeper significance which was only fully appreciated several centuries later by the mathematicians of India. The Hindus recognised that zero had an independent existence beyond the mere spacing role among the other numbers – zero was a number in its own right. It represented a quantity of nothing. For the first time the abstract concept of nothingness had been given a tangible symbolic representation.

This may seem a trivial step forward to the modern reader, but the deeper meaning of the zero symbol had been ignored by all the ancient Greek philosophers, including Aristotle. He had argued that the number zero should be outlawed because it disrupted the consistency of the other numbers – dividing any ordinary number by zero led to an incomprehensible result. By the sixth century the Indian mathematicians no longer brushed this problem under the rug, and the seventh-century scholar Brahmagupta was sophisticated enough to use division by zero as a definition for infinity.

While Europe had abandoned the noble search for truth, India and Arabia were consolidating the knowledge which had been smuggled out of the embers of Alexandria and were reinterpreting it in a new and more eloquent language. As well as adding zero to the mathematical vocabulary, they replaced the primitive Greek symbols and cumbersome Roman numerals with the counting system which has now been universally adopted. Once again, this might seem like an absurdly humble step forward, but try multiplying CLV by DCI and you will appreciate the significance of the breakthrough. The equivalent task of multiplying 155 by 601 is a good deal simpler. The growth of any discipline depends on the ability to communicate and develop ideas, and this in turn relies on a language which is sufficiently detailed and flexible. The ideas of Pythagoras and Euclid were no less elegant for their awkward expression, but translated into the symbols of Arabia they would blossom and give fruit to newer and richer concepts.

In the tenth century the French scholar Gerbert of Aurillac learnt the new counting system from the Moors of Spain and through his teaching positions at churches and schools throughout Europe he was able to introduce the new system to the West.

In 999 he was elected Pope Sylvester II, an appointment which allowed him to further encourage the use of Indo-Arabic numerals. Although the efficiency of the system revolutionised accounting and was rapidly adopted by merchants, it did little to inspire a revival in European mathematics.

The vital turning point for Western mathematics occurred in 1453 when the Turks ransacked Constantinople. During the intervening years the manuscripts which had survived the desecration of Alexandria had congregated in Constantinople, but once again they were threatened with destruction. Byzantine scholars fled westward with whatever texts they could preserve. Having survived the onslaught of Caesar, Bishop Theophilus, Caliph Omar and now the Turks, a few precious volumes of the *Arithmetica* made their way back to Europe. Diophantus was destined for the desk of Pierre de Fermat.

Birth of a Riddle

Fermat's judicial responsibilities occupied a great deal of his time, but what little leisure he had was devoted entirely to mathematics. This was partly because judges in seventeenth-century France were discouraged from socialising on the grounds that friends and acquaintances might one day be called before the court. Fraternising with the locals would only lead to favouritism. Isolated from the rest of Toulouse's high society, Fermat could concentrate on his hobby.

There is no record of Fermat ever being inspired by a mathematical tutor; instead it was a copy of the *Arithmetica* which became his mentor. The *Arithmetica* sought to describe the theory of numbers, as it was in Diophantus' time, via a series of problems and solutions. In effect Diophantus was presenting Fermat with one thousand years worth of mathematical understanding. In one book Fermat could find the entire knowledge of numbers as constructed by the likes of Pythagoras and Euclid. The theory of numbers had stood still ever since the barbaric burning of Alexandria, but now Fermat was ready to resume study of the most fundamental of mathematical disciplines.

The *Arithmetica* which inspired Fermat was a Latin translation made by Claude Gaspar Bachet de Méziriac, reputedly the most learned man in all of France. As well as being a brilliant linguist, poet and classics scholar, Bachet had a passion for mathematical puzzles. His first publication was a compilation of puzzles entitled *Problemes plaisans et délectables qui se font par les nombres*, which included river-crossing problems, a liquid-pouring problem and several think-of-a-number tricks. One of the questions posed was a problem about weights:

What is the least number of weights that can be used on a set of scales to weigh any whole number of kilograms from 1 to 40?

Bachet had a cunning solution which shows that it is possible to achieve this task with only four weights. His solution is given in Appendix 4.

Although he was merely a mathematical dilettante, Bachet's interest in puzzles was enough for him to realise that Diophantus' list of problems were on a higher plane and worthy of deeper study. He set himself the task of translating Diophantus' opus and publishing it so that the techniques of the Greeks could be rekindled. It is important to realise that vast quantities of ancient mathematical knowledge had been completely forgotten. Higher mathematics was not taught in even the greatest European universities and it is only thanks to the efforts of scholars such as Bachet that so much was revived so rapidly. In 1621 when Bachet published the Latin version of the *Arithmetica*, he was contributing to the second golden age of mathematics.

The *Arithmetica* contains over one hundred problems and for each one Diophantus gives a detailed solution. This level of conscientiousness was not a habit which Fermat ever picked up. Fermat was not interested in writing a textbook for future generations: he merely wanted to satisfy himself that he had solved a problem. While studying Diophantus' problems and solutions, he would be inspired to think of and tackle other related and more subtle questions. Fermat would scribble down whatever was necessary to convince himself that he could see the solution and then he would not bother to write down the remainder of the proof. More often than not he would consign his inspirational jottings to the bin, and then move on to the next problem. Fortunately for us, Bachet's publication of the *Arithmetica* contained generous margins on every page, and sometimes Fermat would hastily write logic and comments in these columns. These marginal notes would become an invaluable, if somewhat scanty, record of Fermat's most brilliant calculations.

One of Fermat's discoveries concerned the so-called *friendly numbers*, or *amicable numbers*, closely related to the perfect numbers which had fascinated Pythagoras two thousand years earlier. Friendly numbers are pairs of numbers such that each number is the sum of the divisors of the other number. The Pythagoreans made the extraordinary discovery that 220 and 284 are friendly numbers. The divisors of 220 are 1, 2, 4, 5, 10, 11, 20, 22, 44, 55, 110, and the sum of all these is 284. On the other hand the divisors of 284 are 1, 2, 4, 71, 142, and the sum of all these is 220.

The pair 220 and 284 was said to be symbolic of friendship. Martin Gardner's book *Mathematical Magic Show* tells of talismans sold in the Middle Ages which were inscribed with these numbers on the grounds that wearing the charms would promote love. An Arab numerologist documents the practice of carving 220 on one fruit and 284 on another, and then eating the first one and offering the second one to a lover as a form of mathematical aphrodisiac. Early theologians noted that in Genesis Jacob gave 220 goats to Esau. They believed that the number of goats, one half of a friendly pair, was an expression of Jacob's love for Esau.

No other friendly numbers were identified until 1636 when Fermat discovered the pair 17,296 and 18,416. Although not a profound discovery, it demonstrates Fermat's familiarity with numbers and his love of playing with them. Fermat started a fad for finding friendly numbers; Descartes discovered a third pair (9,363,584 and 9,437,056) and Leonhard Euler went on to list sixty-two amicable pairs. Curiously they had all overlooked a much smaller pair of friendly numbers. In 1866 a sixteen-year-old Italian, Nicolò Paganini, discovered the pair 1,184 and 1,210.

During the twentieth century mathematicians have extended the idea further and have searched for so-called 'sociable' numbers, three or more numbers which form a closed loop. For example, in the loop of 5 numbers (12,496; 14,288; 15,472;14,536; 14,264) the divisors of the first number add up to the second, the divisors of the second add to the third, the divisors of the third add up to the fourth, the divisors of the fourth add up to the fifth, and the divisors of the fifth add up to the first.

Although discovering a new pair of friendly numbers made Fermat something of a celebrity, his reputation was truly confirmed thanks to a series of mathematical challenges. For example, Fermat noticed that 26 is sandwiched between 25 and 27, one of which is a square number ($25 = 5^2 = 5 \times 5$) and the other is a cube number ($27 = 3^3 = 3 \times 3 \times 3$). He searched for other numbers sandwiched between a square and a cube but failed to find any, and suspected that 26 might be unique. After days of strenuous effort he managed to construct an elaborate argument which proved without any doubt that 26 is indeed the only number between a square and a cube. His step-by-step logical proof established that no other numbers could fulfil this criterion.

Fermat announced this unique property of 26 to the mathematical community, and then challenged them to prove that this was the case. He openly admitted that he himself had a proof; the question was, however, did others have the ingenuity to match it? Despite the simplicity of the claim the proof is fiendishly complicated, and Fermat took particular delight in taunting the English mathematicians Wallis and Digby, who eventually had to admit defeat. Ultimately Fermat's greatest claim to fame would turn out to be another challenge to the rest of the world. However, it would be an accidental riddle which was never intended for public discussion.

The Marginal Note

While studying Book II of the *Arithmetica* Fermat came upon a whole series of observations, problems and solutions which concerned Pythagoras' theorem and Pythagorean triples. For instance, Diophantus discussed the existence of particular triples which formed so-called 'limping triangles', ones in which the two shorter legs x and y differ only by one (e.g. $x = 20, y = 21, z = 29$ and $20^2 + 21^2 = 29^2$).

Fermat was struck by the variety and sheer quantity of Pythagorean triples. He was aware that centuries earlier Euclid had stated a proof, outlined in Appendix 5, which demonstrated that, in fact, there are an infinite number of Pythagorean triples. Fermat must have gazed at Diophantus' detailed exposition of Pythagorean triples and wondered what there was to add to the subject. As he stared at the page he began to play with Pythagoras' equation, trying to discover something which had evaded the Greeks. Suddenly, in a moment of genius which would immortalise the Prince of Amateurs, he created an equation which, though very similar to Pythagoras' equation, had no solutions at all. This was the equation which the ten-year-old Andrew Wiles read about in the Milton Road Library.

Instead of considering the equation

$$x^2 + y^2 = z^2,$$

Fermat was contemplating a variant of Pythagoras' creation:

$$x^3 + y^3 = z^3.$$

As mentioned in the last chapter, Fermat had merely changed the power from 2 to 3, the square to a cube, but his new equation apparently had no whole number solutions whatsoever. Trial and error soon shows the difficulty of finding two cubed numbers which add together to make another cubed number. Could it really be the case that this minor modification turns Pythagoras' equation, one with an infinite number of solutions, into an equation with no solutions?

He altered the equation further by changing the power to numbers bigger than 3, and discovered that finding a solution to each of these equations was equally difficult. According to Fermat there appeared to be no three numbers which would perfectly fit the equation

$$x^n + y^n = z^n, \quad \text{where } n \text{ represents } 3, 4, 5, \ldots$$

In the margin of his *Arithmetica*, next to Problem 8, he made a note of his observation:

Cubem autem in duos cubos, aut quadratoquadratum in duos quadratoquadratos, et generaliter nullam in infinitum ultra quadratum potestatem in duos eiusdem nominis fas est dividere.

It is impossible for a cube to be written as a sum of two cubes or a fourth power to be written as the sum of two fourth powers or, in general, for any number which is a power greater than the second to be written as a sum of two like powers.

Among all the possible numbers there seemed to be no reason why at least one set of solutions could not be found, yet Fermat stated that nowhere in the infinite universe of numbers was there a 'Fermatean triple'. It was an extraordinary claim, but one which Fermat believed he could prove. After the first marginal note outlining the theory, the mischievous genius jotted down an additional comment which would haunt generations of mathematicians:

Cuius rei demonstrationem mirabilem sane detexi hanc marginis exiguitas non caperet.

I have a truly marvellous demonstration of this proposition which this margin is too narrow to contain.

This was Fermat at his most infuriating. His own words suggest that he was particularly pleased with this 'truly marvellous' proof, but he had no intention of bothering to write out the detail of the argument, never mind publishing it. He never told anyone about his proof, and yet despite his combination of indolence and modesty Fermat's Last Theorem, as it would later be called, would become famous around the world for centuries to come.

The Last Theorem Published at Last

Fermat's notorious discovery happened early in his mathematical career, in around 1637. Some thirty years later, while carrying out his judicial duties in the town of Castres, Fermat was taken seriously ill. On 9 January 1665, he signed his last *arrêt*, and three days later he died. Still isolated from the Parisian school of mathematics and not necessarily fondly remembered by his frustrated correspondents, Fermat's discoveries were at risk of being lost forever. Fortunately Fermat's eldest son, Clément-Samuel, who appreciated the significance of his father's hobby, was determined that his discoveries should not be lost to the world. It is thanks to his efforts that we know anything at all about Fermat's remarkable breakthroughs in number theory and, in particular, if it were not for Clément-Samuel, the enigma known as Fermat's Last Theorem would have died with its creator.

Clément-Samuel spent five years collecting his father's notes and letters, and examining the jottings in the margins of his copy of the *Arithmetica*. The marginal note referring to Fermat's Last Theorem was just one of many inspirational thoughts scribbled in the book, and Clément-Samuel undertook to publish these annotations in a special edition of the *Arithmetica*. In 1670 at Toulouse he brought out *Diophantus' Arithmetica Containing Observations by P. de Fermat*. Alongside Bachet's original Greek and Latin translations were forty-eight observations made by Fermat. The observation, shown in Figure 6, was to be the one which would become known as Fermat's Last Theorem.

Once Fermat's *Observations* reached the wider community, it was clear that the letters he had sent to colleagues represented mere morsels from a treasure trove of discovery. His personal notes

DIOPHANTI
ALEXANDRINI
ARITHMETICORVM
LIBRI SEX,
ET DE NVMERIS MVLTANGVLIS
LIBER VNVS.

CVM COMMENTARIIS C. G. BACHETI V. C.
& obseruationibus D. P. de FERMAT Senatoris Tolosani.

Accessit Doctrinæ Analyticæ inuentum nouum, collectum
ex varijs eiusdem D. de FERMAT Epistolis.

OBLOQVITVR NVMERIS SEPTEM DISCRIMINA VOCVM

TOLOSÆ,
Excudebat BERNARDVS BOSC, è Regione Collegij Societatis Iesu.

M. DC. LXX.

The frontispiece of Clément-Samuel Fermat's edition of
Diophantus' *Arithmetica*, published in 1670. This version included
the marginal notes made by his father.

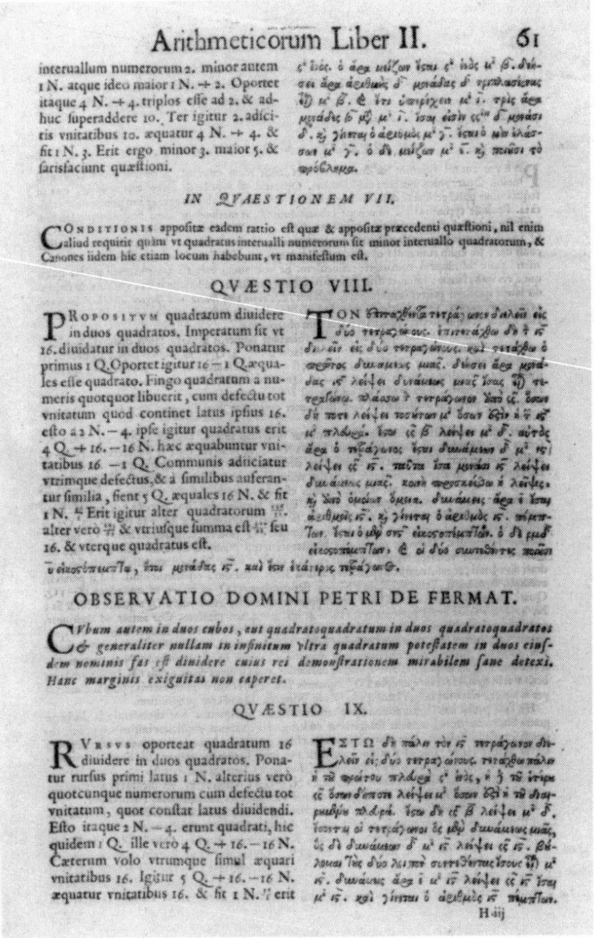

Arithmeticorum Liber II. 61

interuallum numerorum 2. minor autem 1 N. atque ideo maior 1 N. + 2. Oportet itaque 4 N. + 4. triplos esse ad 2. & adhuc superaddere 10. Ter igitur 2. adscitis vnitatibus 10. æquatur 4 N. + 4. & fit 1 N. 3. Erit ergo minor 3. maior 5. & satisfaciunt quæstioni.

IN QVAESTIONEM VII.

CONDITIONIS appositæ eadem ratio est quæ & appositæ præcedenti quæstioni, nil enim aliud requirit quàm vt quadratus interualli numerorum sit minor interuallo quadratorum, & Canones iidem hic etiam locum habebunt, vt manifestum est.

QVÆSTIO VIII.

PROPOSITVM quadratum diuidere in duos quadratos. Imperatum sit vt 16. diuidatur in duos quadratos. Ponatur primus 1 Q. Oportet igitur 16 − 1 Q. æquales esse quadrato. Fingo quadratum a numeris quotquot libuerit, cum defectu tot vnitatum quod continet latus ipsius 16. esto a 2 N. − 4. ipse igitur quadratus erit 4 Q. + 16. − 16 N. hæc æquabuntur vnitatibus 16. − 1 Q. Communis adiiciatur vtrimque defectus, & a similibus auferantur similia, fient 5 Q. æquales 16 N. & fit 1 N. $\frac{16}{5}$ Erit igitur alter quadratorum $\frac{256}{25}$. alter verò $\frac{144}{25}$ & vtriusque summa est $\frac{400}{25}$ seu 16. & vterque quadratus est.

OBSERVATIO DOMINI PETRI DE FERMAT.

CVbum autem in duos cubos, aut quadratoquadratum in duos quadratoquadratos & generaliter nullam in infinitum vltra quadratum potestatem in duos eiusdem nominis fas est diuidere cuius rei demonstrationem mirabilem sane detexi. Hanc marginis exiguitas non caperet.

QVÆSTIO IX.

RVRSVS oporteat quadratum 16 diuidere in duos quadratos. Ponatur rursus primi latus 1 N. alterius verò quotcunque numerorum cum defectu tot vnitatum, quot constat latus diuidendi. Esto itaque 2 N. − 4. erunt quadrati, hic quidem 1 Q. ille verò 4 Q. + 16. − 16 N. Cæterum volo vtrumque simul æquari vnitatibus 16. Igitur 5 Q. + 16. − 16 N. æquatur vnitatibus 16. & fit 1 N. $\frac{16}{5}$ erit

H iij

Figure 6. The page containing Pierre de Fermat's notorious observation.

contained a whole series of theorems. Unfortunately these were accompanied either with no explanation at all or with only a slight hint of the underlying proof. There were just enough tantalising glimpses of logic to leave mathematicians in no doubt that Fermat had proofs, but filling in the details was left as a challenge for them to take up.

Leonhard Euler, one of the greatest mathematicians of the eighteenth century, attempted to prove one of Fermat's most elegant observations, a theorem concerning prime numbers. A prime number is one which has no divisors – no number will divide into it without a remainder, except for 1 and the number itself. For instance, 13 is a prime number, but 14 is not. Nothing will divide into 13 perfectly, but 2 and 7 will divide into 14. All prime numbers can be put into two categories; those which equal $4n + 1$ and those which equal $4n - 1$, where n equals some number. So 13 is in the former group ($4 \times 3 + 1$), whereas 19 is in the latter group ($4 \times 5 - 1$). Fermat's prime theorem claimed that the first type of primes were always the sum of two squares ($13 = 2^2 + 3^2$), whereas the second type could never be written in this way ($19 = ?^2 + ?^2$). This property of primes is beautifully simple, but trying to prove that it is true for every single prime number turns out to be remarkably difficult. For Fermat it was just one of many private proofs. The challenge for Euler was to rediscover Fermat's proof. Eventually in 1749, after seven years work and almost a century after Fermat's death, Euler succeeded in proving this prime number theorem.

Fermat's panoply of theorems ranged from the fundamental to the simply amusing. Mathematicians rank the importance of theorems according to their impact on the rest of mathematics. First, a theorem is considered important if it has a universal truth, that is to say, if it applies to an entire group of numbers. In the case of the prime number theorem, it is true not for just some prime numbers, but for all prime numbers. Second, theorems should reveal some deeper underlying truth about the relationship between numbers. A theorem can be the springboard for generating a whole host of other theorems, even inspiring the development of whole new branches of mathematics. Finally, a theorem is important if entire areas of research can be hindered for the lack of one logical link. Many mathematicians have cried themselves to sleep knowing that they could achieve a major result if only they could establish one missing link in their chain of logic.

Because mathematicians employ theorems as stepping stones to other results, it was essential that every single one of Fermat's theorems be proved. Just because Fermat said he had a proof of a theorem it could not be accepted at face value. Before it could be used, each theorem had to be proved with ruthless rigour, otherwise the consequences could have been disastrous. For example, imagine that mathematicians had accepted one of Fermat's theorems. It would then be incorporated as a single element in a whole series of other larger proofs. In due course these larger proofs would be incorporated into even larger proofs, and so on. Ultimately hundreds of theorems could come to rely on the truth of the original unchecked theorem. However, what if Fermat had made a mistake and the unchecked theorem was in fact flawed? All these other theorems which incorporated it would also be flawed, and vast areas of mathematics would collapse. Theorems are the foundations of mathematics, because once their truth has been established other theorems can safely be built on top of them. Unsubstantiated ideas are infinitely less valuable and are referred to as conjectures. Any logic which relies on a conjecture is itself a conjecture.

Fermat said he had a proof for every one of his observations, so for him they were theorems. However, until the community at large could rediscover the individual proofs each one could only be considered a conjecture. In fact for the last 350 years Fermat's Last Theorem should more accurately have been referred to as Fermat's Last Conjecture.

As the centuries passed, all his other observations were proved one by one, but Fermat's Last Theorem stubbornly refused to give in so easily. In fact, it is called the 'Last' Theorem because it remains the last one of the observations to be proved. Three centuries of effort failed to find a proof, and this led to its notoriety as the most demanding riddle in mathematics. However, this acknowledged difficulty does not necessarily mean that Fermat's Last Theorem is an important theorem in the ways described earlier. The Last Theorem, at least until very recently, seemed to fail to fulfil several criteria – it seemed that proving it would not lead to anything profound, it would not give any particularly deep insight about numbers, and it would not help prove any other conjectures.

The fame of Fermat's Last Theorem comes solely from the sheer difficulty of proving it. An extra sparkle is added by the fact that the Prince of Amateurs said that he could prove this theorem which has since baffled generations of professional mathematicians. Fermat's offhand comments in the margin of his copy of the *Arithmetica* were read as a challenge to the world. He had proved the Last Theorem: the question was, could any mathematician match his brilliance?

G.H. Hardy had a whimsical sense of humour and dreamt up what could have been an equally frustrating legacy. Hardy's challenge was in the form of an insurance policy to help him cope with his fear of travelling on ships. If he ever had to journey across the sea he would first send a telegram to a colleague saying:

HAVE SOLVED RIEMANN HYPOTHESIS STOP
WILL GIVE DETAILS UPON RETURN STOP

The Riemann hypothesis is a problem which has plagued mathematicians since the nineteenth century. Hardy's logic was that God would never allow him to drown because it would leave mathematicians haunted by a second terrible phantom.

Fermat's Last Theorem is a problem of immense difficulty, and yet it can be stated in a form that a schoolchild can understand. There can be no problem in physics, chemistry or biology which can be so simply and unambiguously stated and which has remained unsolved for so long. In his book *The Last Problem*, E.T. Bell wrote that civilisation would probably come to an end before Fermat's Last Theorem could be solved. Proving Fermat's Last Theorem has become the most valuable prize in number theory, and not surprisingly it has led to some of the most exciting episodes in the history of mathematics. The search for a proof of Fermat's Last Theorem has involved the greatest minds on the planet, huge rewards, suicidal despair and duelling at dawn.

The riddle's status has gone beyond the closed world of mathematics. In 1958 it even made its way into a Faustian tale. An anthology entitled *Deals with the Devil* contains a short story written by Arthur Poges. In 'The Devil and Simon Flagg' the Devil asks Simon Flagg to set him a question. If the Devil succeeds in answering it within twenty-four hours then he takes Simon's soul, but if he fails then he must give Simon $100,000. Simon poses the question: 'Is Fermat's Last Theorem correct?' The Devil disappears and whizzes around the world to absorb every piece of mathematics that has ever been created. The following day he returns and admits defeat:

'You win, Simon,' he said, almost in a whisper, eyeing him with ungrudging respect. 'Not even I can learn enough mathematics in such a short time for so difficult a problem. The more I got into it the worse it became. Non-unique factoring, ideals – Bah! Do you know,' the Devil confided, 'not even the best mathematicians on other planets – all far ahead of yours – have solved it? Why, there's a chap on Saturn – he looks something like a mushroom on stilts – who solves partial differential equations mentally; and even he's given up.'

Leonhard Euler

3

A Mathematical Disgrace

Mathematics is not a careful march down a well-cleared highway, but a journey into a strange wilderness, where the explorers often get lost. Rigour should be a signal to the historian that the maps have been made, and the real explorers have gone elsewhere.

W.S. Anglin

'Since I first met Fermat's Last Theorem as a child it's been my greatest passion,' recalls Andrew Wiles, in a hesitant voice which conveys the emotion he feels about the problem. 'I'd found this problem which had been unsolved for three hundred years. I don't think many of my schoolfriends caught the mathematics bug, so I didn't discuss it with my contemporaries. But I did have a teacher who had done research in mathematics and he gave me a book about number theory that gave me some clues about how to start tackling it. To begin with I worked on the assumption that Fermat didn't know very much more mathematics than I would have known. I tried to find his lost solution by using the kind of methods he might have used.'

Wiles was a child full of innocence and ambition, who saw an opportunity to succeed where generations of mathematicians had failed. To others this might have seemed like a foolhardy dream but young Andrew was right in thinking that he, a twentieth-

century schoolboy, knew as much mathematics as Pierre de
Fermat, a genius of the seventeenth century. Perhaps in his naïvety
he would stumble upon a proof which other more sophisticated
minds had missed.

Despite his enthusiasm every calculation resulted in a dead end.
Having racked his brains and sifted through his schoolbooks he
was achieving nothing. After a year of failure he changed his
strategy and decided that he might be able to learn something from
the mistakes of other more eminent mathematicians. 'Fermat's
Last Theorem has this incredible romantic history to it. Many
people have thought about it, and the more that great mathematicians in the past have tried and failed to solve the problem, the
more of a challenge and the more of a mystery it's become. Many
mathematicians had tried it in so many different ways in the eighteenth and nineteenth centuries, and so as a teenager I decided that
I ought to study those methods and try to understand what they'd
been doing.'

Young Wiles examined the approaches of everyone who had
ever made a serious attempt to prove Fermat's Last Theorem. He
began by studying the work of the most prolific mathematician in
history and the first one to make a breakthrough in the battle
against Fermat.

The Mathematical Cyclops

Creating mathematics is a painful and mysterious experience.
Often the object of the proof is clear, but the route is shrouded in
fog, and the mathematician stumbles through a calculation, terri-
fied that each step might be taking the argument in completely the
wrong direction. Additionally there is the fear that no route exists.

A mathematician may believe that a statement is true, and spend years trying to prove that it is indeed true, when all along it is actually false. The mathematician has effectively been attempting to prove the impossible.

In the entire history of the subject only a handful of mathematicians appear to have avoided the self-doubt which intimidates their colleagues. Perhaps the most notable example of such a mathematician was the eighteenth-century genius Leonhard Euler, and it was he who made the first breakthrough towards proving Fermat's Last Theorem. Euler had such an incredible intuition and vast memory that it was said he could map out the entire bulk of a calculation in his head without having to put pen to paper. Across Europe he was referred to as 'analysis incarnate', and the French academician François Arago said, 'Euler calculated without apparent effort as men breathe, or as eagles sustain themselves in the wind.'

Leonhard Euler was born in Basle in 1707, the son of a Calvinist pastor, Paul Euler. Although the young Euler showed a prodigious talent for mathematics, his father was determined that he should study theology and pursue a career in the Church. Leonhard dutifully obeyed and studied theology and Hebrew at the University of Basle.

Fortunately for Euler the town of Basle was also home to the eminent Bernoulli clan. The Bernoullis could easily claim to be the most mathematical of families, creating eight of Europe's most outstanding minds within only three generations – some have said that the Bernoulli family was to mathematics what the Bach family was to music. Their fame spread beyond the mathematical community and one particular legend typifies the profile of the family. Daniel Bernoulli was once travelling across Europe and had struck up a conversation with a stranger. After a while he modestly introduced himself: 'I am Daniel Bernoulli.' 'And I,' said his companion sarcastically, 'am Isaac Newton.' Daniel fondly recalled this incident on several occasions, considering it the most sincere tribute he had ever received.

Daniel and Nikolaus Bernoulli were close friends of Leonhard Euler, and they realised that the most brilliant of mathematicians was being turned into the most mediocre of theologians. They appealed to Paul Euler and requested that Leonhard be allowed to forsake the cloth in favour of numbers. Euler senior had in the past been taught mathematics by Bernoulli senior, Jakob, and had a tremendous respect for the family. Reluctantly he accepted that his son had been born to calculate, not preach.

Leonhard Euler soon left Switzerland for the palaces of Berlin and St Petersburg, where he was to spend the bulk of his creative years. During the era of Fermat, mathematicians were considered amateur number-jugglers, but by the eighteenth century they were treated as professional problem-solvers. The culture of numbers had changed dramatically, and this was partly a consequence of Sir Isaac Newton and his scientific calculations.

Newton believed that mathematicians were wasting their time teasing each other with pointless riddles. Instead he would apply mathematics to the physical world and calculate everything from the orbits of the planets to the trajectories of cannon-balls. By the time Newton died, in 1727, Europe had undergone a scientific revolution, and in the same year Euler published his first paper. Although the paper contained elegant and innovative mathematics, it was primarily aimed at describing a solution to a technical problem regarding the masting of ships.

The European powers were not interested in using mathematics to explore esoteric and abstract concepts; instead they wanted to exploit mathematics to solve practical problems, and they competed to employ the best minds. Euler began his career with the Czars, before being invited to the Berlin Academy by Frederick the Great of Prussia. Eventually he returned to Russia, under the rule of Catherine the Great, where he spent his final years. During his career he tackled a multitude of problems, ranging from navigation to finance, and from acoustics to irrigation. The practical world of problem-solving did not dull Euler's mathematical ability. Instead tackling each new task would inspire him to create innovative and ingenious mathematics. His single-minded passion drove him to write several papers in a single day, and it is said that between the first and second calls for dinner he would attempt to dash off a complete calculation worthy of publication. Not a moment was wasted and even when he was cradling an infant in one hand Euler would be outlining a proof with the other.

One of Euler's greatest achievements was the development of the algorithmic method. The point of Euler's algorithms was to tackle apparently impossible problems. One such problem was predicting the phases of the moon far into the future with high accuracy – information which could be used to draw up vital navigation tables. Newton had already shown that it is relatively easy to predict the orbit of one body around another, but in the case of the moon the situation is not so simple. The moon orbits the earth, but there is a third body, the sun, which complicates matters enormously. While the earth and moon attract each other, the sun perturbs the position of the earth and has a knock-on effect on the orbit of the moon. Equations could be used to pin down the effect of any two of the bodies, but eighteenth-century mathematicians could not incorporate the third body into their calculations. Even today it is impossible to predict the exact solution to the so-called 'three-body problem'.

Euler realised that mariners did not need to know the phase of the moon with absolute accuracy, only with enough precision to locate their own position to within a few nautical miles. Consequently Euler developed a recipe for generating an imperfect but sufficiently accurate solution. The recipe, known as an algorithm, worked by first obtaining a rough-and-ready result, which could then be fed back into the algorithm to generate a more refined result. This refined result could then be fed back into the algorithm to generate an even more accurate result, and so on. A hundred or so iterations later Euler was able to provide a position for the moon which was accurate enough for the purposes of the navy. He gave his algorithm to the British Admiralty and in return they rewarded him with a prize of £300.

Euler earned a reputation for being able to solve any problem which was posed, a talent which seemed to extend even beyond the realm of science. During his stint at the court of Catherine the Great he encountered the great French philosopher Denis Diderot. Diderot was a committed atheist and would spend his days converting the Russians to atheism. This infuriated Catherine, who asked Euler to put a stop to the efforts of the godless Frenchman.

Euler gave the matter some thought and claimed that he had an algebraic proof for the existence of God. Catherine the Great invited Euler and Diderot to the palace and gathered together her courtiers to listen to the theological debate. Euler stood before the audience and announced:

$$\text{`Sir,} \quad \frac{a + b^n}{n} = x, \quad \text{hence God exists; reply!'}$$

With no great understanding of algebra, Diderot was unable to argue against the greatest mathematician in Europe and was left speechless. Humiliated, he left St Petersburg and returned to Paris.

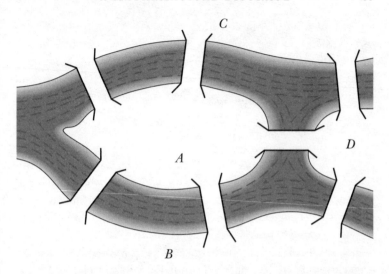

Figure 7. The River Pregel divides the town of Königsberg into four separate parts, *A*, *B*, *C* and *D*. Seven bridges connect the various parts of the town, and a local riddle asked if it was possible to make a journey such that each bridge is crossed once, and only once.

In his absence, Euler continued to enjoy his return to theological study and published several other mock proofs concerning the nature of God and the human spirit.

A more valid problem which also appealed to Euler's whimsical nature concerned the Prussian city of Königsberg, now known as the Russian city of Kaliningrad. The city is built on the banks of the river Pregel and consists of four separate quarters connected by seven bridges. Figure 7 shows the layout of the city. Some of the more curious residents of Königsberg wondered if it was possible to plot a journey across all seven bridges without having to stroll

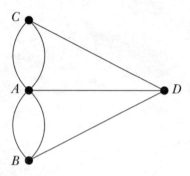

Figure 8. A simplified representation of the bridges of Königsberg.

across any bridge more than once. The citizens of Königsberg tried various routes but each one ended in failure. Euler also failed to find a successful route, but he was successful in explaining why making such a journey was impossible.

Euler began with a plan of the city, and from it he generated a simplified representation in which the sections of land were reduced to points and bridges were replaced by lines, as shown in Figure 8. He then argued that, in general, in order to make a successful journey (i.e. crossing all bridges only once) a point should be connected to an even number of lines. This is because in the middle of a journey when the traveller passes through a land mass, he or she must enter via one bridge and then leave via a different bridge. There are only two exceptions to this rule – when a traveller either begins or ends the journey. At the start of the journey the traveller leaves a land mass and requires only a single bridge to exit, and at the end of the journey the traveller arrives at a land mass and requires only a single bridge to enter. If the journey begins and ends in different locations, then these two land masses are allowed to have an odd number of bridges. But if the journey begins and ends in the same place, then this point, like all the other points, must have an even number of bridges.

So, in general, Euler concluded that, for any network of bridges, it is only possible to make a complete journey crossing each bridge only once if all the landmasses have an even number of bridges, or exactly two land masses have an odd number of bridges. In the case of Königsberg there are four land masses in total and all of them are connected to an odd number of bridges – three points have three bridges, and one has five bridges. Euler had been able to explain why it was impossible to cross each one of Königsberg's bridges once and only once, and furthermore he had generated a rule which could be applied to any network of bridges in any city in the world. The argument is beautifully simple, and was perhaps just the sort of logical problem that Euler dashed off before dinner.

The Königsberg bridge puzzle is a so-called network problem in applied mathematics, but it inspired Euler to consider more abstract networks. He went on to discover a fundamental truth about all networks, the so-called *network formula*, which he could prove with just a handful of logical steps. The network formula shows an eternal relationship between the three properties which describe any network:

$$V + R - L = 1,$$

where

  $V =$ the number of vertices (intersections) in the network,

  $L =$ the number of lines in the network,

  $R =$ the number of regions (enclosed areas) in the network.

Euler claimed that for any network one could add the number of vertices and regions and subtract the number of lines and the total

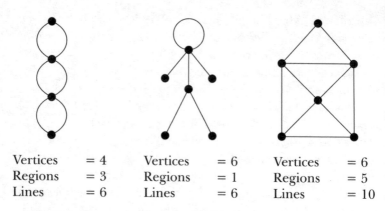

Vertices	= 4	Vertices	= 6	Vertices	= 6
Regions	= 3	Regions	= 1	Regions	= 5
Lines	= 6	Lines	= 6	Lines	= 10

Figure 9. All conceivable networks obey Euler's network formula.

would always be 1. For example, all the networks in Figure 9 obey the rule.

It is possible to imagine testing this formula on a whole series of networks and if it turned out to be true on each occasion it would be tempting to assume that the formula is true for all networks. Although this might be enough evidence for a scientific theory, it is inadequate to justify a mathematical theorem. The only way to show that the formula works for every possible network is to construct a foolproof argument, which is exactly what Euler did.

Euler began by considering the simplest network of all, i.e. a single vertex as shown in Figure 10(a). For this network the formula is clearly true: there is one vertex, and no lines or regions, and therefore

$$V + R - L = 1 + 0 - 0 = 1.$$

Euler then considered what would happen if he added something to this simplest of all networks. Any extension to the single vertex

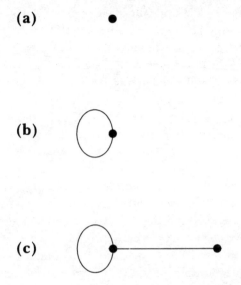

Figure 10. Euler proved his network formula by showing that it was true for the simplest network, and then demonstrating that the formula would remain true whatever extensions were added to the single vertex.

requires the addition of a line. The line can either connect the existing vertex to itself, or it can connect the existing vertex to a new vertex.

First, let us look at connecting the vertex to itself with this additional line. As shown in Figure 10(b), when the line is added, this also results in a new region. Therefore the network formula remains true because the extra region (+1) cancels the extra line (−1). If further lines are added in this way the network formula will still remain true because each new line will create a new region.

Second, let us look at using the line to connect the original vertex to a new vertex, as shown in Figure 10(c). Once again the network formula remains true because the extra vertex ($+1$) cancels the extra line (-1). If further lines are added in this way, the network formula will still remain true because each new line will create a new vertex.

This was all that Euler required for his proof. He argued that the network formula was true for the simplest of all networks, the single vertex. Furthermore, all other networks, no matter how complicated, can be constructed from the simplest network by adding lines one at a time. Each time a new line is added the network formula will remain true because either a new vertex or a new region will always be added and this will have a compensating effect. Euler had developed a simple but powerful strategy. He proved that the formula is true for the most basic network, a single vertex, and then he demonstrated that any operation which complicated the network would continue to conserve the validity of the formula. Therefore the formula is true for the infinity of all possible networks.

When Euler first encountered Fermat's Last Theorem, he must have hoped that he could solve it by adopting a similar strategy. The Last Theorem and the network formula come from very different areas of mathematics but they have one thing in common, which is that both say something about an infinite number of objects. The network formula says that for the infinite number of networks that exist the number of vertices and regions less the number of lines always equals 1. Fermat's Last Theorem claims that for an infinite number of equations there are no whole number solutions. Recall that Fermat stated that there are no whole number solutions to the following equation:

$$x^n + y^n = z^n, \quad \text{where } n \text{ is any number greater than 2.}$$

This equation represents an infinite set of equations:

$$x^3 + y^3 = z^3,$$
$$x^4 + y^4 = z^4,$$
$$x^5 + y^5 = z^5,$$
$$x^6 + y^6 = z^6,$$
$$x^7 + y^7 = z^7,$$
$$\vdots$$

Euler wondered if he could prove that one of the equations had no solutions and then extrapolate the result to all the remaining equations, in the same way he had proved his network formula for all networks by generalising it from the simplest case, the single vertex.

Euler's task was given a head start when he discovered a clue hidden in Fermat's jottings. Although Fermat never wrote down a proof for the Last Theorem, he did cryptically describe a proof for the specific case $n = 4$ elsewhere in his copy of the *Arithmetica* and incorporated it into the proof of a completely different problem. Even though this is the most complete calculation he ever committed to paper, the details are still sketchy and vague, and Fermat concludes the proof by saying that lack of time and paper prevent him from giving a fuller explanation. Despite the lack of detail in Fermat's scribbles, they clearly illustrate a particular form of proof by contradiction known as the *method of infinite descent*.

In order to prove that there were no solutions to the equation $x^4 + y^4 = z^4$, Fermat began by assuming that there was a hypothetical solution

$$x = X_1, \qquad y = Y_1, \qquad z = Z_1.$$

By examining the properties of (X_1, Y_1, Z_1), Fermat could demonstrate that if this hypothetical solution did exist then there would have to be a smaller solution (X_2, Y_2, Z_2). Then, by examining this new solution, Fermat could show there would be an even smaller solution (X_3, Y_3, Z_3), and so on.

Fermat had discovered a descending staircase of solutions, which theoretically would continue forever, generating ever-smaller numbers. However, x, y and z must be whole numbers, and so the never-ending staircase is impossible because there must be a smallest possible solution. This contradiction proves that the initial assumption that there is a solution (X_1, Y_1, Z_1) must be false. Using the method of infinite descent Fermat had demonstrated that it is forbidden for the equation with $n = 4$ to have any solutions, because otherwise the consequences would be absurd.

Euler tried to use this as a starting point for constructing a general proof for all the other equations. As well as building up to $n = $ infinity, he would also have to build down to $n = 3$ and it was this single downward step which he attempted first. On 4 August 1753 Euler announced in a letter to the Prussian mathematician Christian Goldbach that he had adapted Fermat's method of infinite descent and successfully proved the case for $n = 3$. After a hundred years this was the first time anybody had succeeded in making any progress towards meeting Fermat's challenge.

In order to extend Fermat's proof from $n = 4$ to cover the case $n = 3$ Euler had to incorporate the bizarre concept of a so-called *imaginary number*, an entity which had been discovered by European mathematicians in the sixteenth century. It is strange to think of new numbers being 'discovered', but this is mainly because we are so familiar with the numbers we commonly use that we forget that there was a time when some of these numbers were not known. Negative numbers, fractions and irrational numbers all had to be discovered and the motivation in each case was to answer otherwise unanswerable questions.

The history of numbers begins with the simple counting numbers (1, 2, 3, …) otherwise known as natural numbers. These numbers are perfectly satisfactory for adding together simple whole quantities, such as sheep or gold coins, to achieve a total number which is also a whole quantity. As well as addition, the other simple operation of multiplication also acts upon whole numbers to generate other whole numbers. However, the operation of division throws up an awkward problem. While 8 divided by 2 equals 4, we find that 2 divided by 8 equals $\frac{1}{4}$. The result of the latter division is not a whole number but a fraction.

Division is a simple operation performed on natural numbers which requires us to look beyond the natural numbers in order to obtain the answer. It is unthinkable for mathematicians not, in theory at least, to be able to answer every single question, and this necessity is called *completeness*. There are certain questions concerning natural numbers which would be unanswerable without resorting to fractions. Mathematicians express this by saying that fractions are necessary for completeness.

It is this need for completeness which led the Hindus to discover negative numbers. The Hindus noticed that, while 3 subtracted from 5 was obviously 2, subtracting 5 from 3 was not such a simple matter. The answer was beyond the natural counting numbers, and could only be accommodated by introducing the concept of negative numbers. Some mathematicians did not accept this extension into abstraction and referred to negative numbers as 'absurd' or 'fictitious'. While an accountant could hold one gold coin, or even half a gold coin, it was impossible to hold a negative coin.

The Greeks also had a yearning for completeness and this led them to discover irrational numbers. In Chapter 2 the question

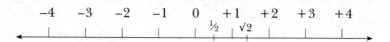

Figure 11. All numbers can be positioned along the number line, which extends to infinity in both directions.

arose, What number is the square root of two, $\sqrt{2}$? The Greeks knew that this number was roughly equal to $\frac{7}{5}$, but when they tried to discover the exact fraction they found that it did not exist. Here was a number which could never be represented as a fraction, but this new type of number was necessary in order to answer a simple question, What is the square root of two? The demand for completeness meant that yet another colony was added to the empire of numbers.

By the Renaissance, mathematicians assumed that they had discovered all the numbers in the universe. All numbers could be thought of as lying on a *number line*, an infinitely long line with zero at the centre, as shown in Figure 11. The whole numbers were spaced equally along the number line, with the positive numbers on the right of zero extending to positive infinity and the negative numbers on the left of zero extending to negative infinity. The fractions occupied the spaces between the whole numbers, and the irrational numbers were interspersed between the fractions.

The number line suggested that completeness had apparently been achieved. All the numbers seemed to be in place, ready to answer all mathematical questions – in any case, there was no more room on the number line for any new numbers. Then during the sixteenth century there were renewed rumblings of disquiet. The Italian mathematician Rafaello Bombelli was studying the square roots of various numbers when he stumbled upon an unanswerable question.

The problem began by asking, What is the square root of one, $\sqrt{1}$? The obvious answer is 1, because $1 \times 1 = 1$. The less obvious answer is -1. A negative number multiplied by another negative number generates a positive number. This means $-1 \times -1 = +1$. So, the square root of $+1$ is both $+1$ and -1. This abundance of answers is fine, but then the question arises, What is the square root of negative one, $\sqrt{-1}$? The problem seems to be intractable. The solution cannot be $+1$ or -1, because the square of both these numbers is $+1$. However, there are no other obvious candidates. At the same time completeness demands that we must be able to answer the question.

The solution for Bombelli was to create a new number, i, called an *imaginary number*, which was simply defined as the solution to the question, *What is the square root of negative one?* This might seem like a cowardly solution to the problem, but it was no different to the way in which negative numbers were introduced. Faced with an otherwise unanswerable question the Hindus merely defined -1 as the solution to the question, *What is zero subtract one?* It is easier to accept the concept of -1 only because we have experience of the analogous concept of 'debt', whereas we have nothing in the real world to underpin the concept of an imaginary number. The seventeenth-century German mathematician Gottfried Leibniz elegantly described the strange nature of the imaginary number: 'The imaginary number is a fine and wonderful recourse of the divine spirit, almost an amphibian between being and non-being.'

Once we have defined i as being the square root of -1, then $2i$ must exist, because this would be the sum of i plus i (as well as being the square root of -4). Similarly $\frac{i}{2}$ must exist because this is the result of dividing i by 2. By performing simple operations it is possible to achieve an imaginary equivalent of every so-called *real number*. There are imaginary counting numbers, imaginary negative numbers, imaginary fractions and imaginary irrationals.

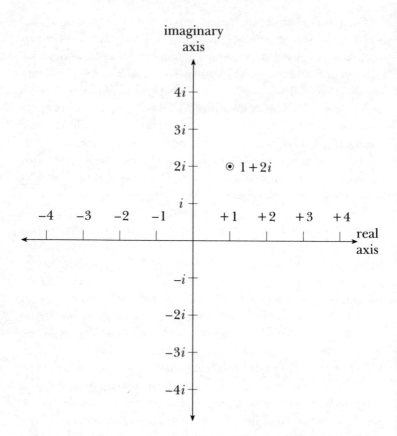

Figure 12. The introduction of an axis for imaginary numbers turns the number line into a number plane. Any combination of real and imaginary numbers has a position on the number plane.

The problem which now arises is that all these imaginary numbers have no natural position along the real number line. Mathematicians resolve this crisis by creating a separate imaginary number line which is perpendicular to the real one, and which crosses at zero, as shown in Figure 12. Numbers are now no longer restricted to a one-dimensional line, but rather they occupy a two-dimensional plane. While pure imaginary or pure real numbers are restricted to their respective lines, combinations of real and imaginary numbers (e.g. $1+2i$), called *complex numbers*, live on the so-called number plane.

What is particularly remarkable is that complex numbers can be used to solve any conceivable equation. For example, in order to calculate $\sqrt{(3 + 4i)}$, mathematicians do not have to resort to inventing a new type of number – the answer turns out to be $2 + i$, another complex number. In other words the imaginary numbers appear to be the final element required to make mathematics complete.

Although the square roots of negative numbers have been referred to as imaginary numbers, mathematicians consider i no more abstract than a negative number or any counting number. In addition, physicists discovered that imaginary numbers provide the best language for describing some real-world phenomena. With a few minor manipulations imaginary numbers turn out to be the ideal way to analyse the natural swinging motion of objects such as pendula. This motion, technically called a sinusoidal oscillation, is found throughout nature, and so imaginary numbers have become an integral part of many physical calculations. Nowadays electrical engineers conjure up i to analyse oscillating currents, and theoretical physicists calculate the consequences of oscillating quantum mechanical wave functions by summoning up the powers of imaginary numbers.

Pure mathematicians have also exploited imaginary numbers, using them to find answers to previously impenetrable problems. Imaginary numbers literally add a new dimension to mathematics, and Euler hoped to exploit this extra degree of freedom to attack Fermat's Last Theorem.

In the past other mathematicians had tried to adapt Fermat's method of infinite descent to work for cases other than $n = 4$, but in every case attempts to stretch the proof only led to gaps in the logic. However, Euler showed that by incorporating the imaginary number, i, into his proof he could plug holes in the proof, and force the method of infinite descent to work for the case $n = 3$.

It was a tremendous achievement, but one which he could not repeat for other cases of Fermat's Last Theorem. Unfortunately Euler's endeavours to make the argument work for the cases up to infinity all ended in failure. The man who created more mathematics than anybody else in history was humbled by Fermat's challenge. His only consolation was that he had made the first breakthrough in the world's hardest problem.

Undaunted by this failure Euler continued to create brilliant mathematics until the day he died, an achievement made all the more remarkable by the fact that during the final years of his career he was totally blind. His loss of sight began in 1735 when the Academy in Paris offered a prize for the solution to an astronomical problem. The problem was so awkward that the mathematical community asked the Academy to allow them several months in which to come up with an answer, but for Euler this was unnecessary. He became obsessed with the task, worked continually for three days and duly won the prize. However, poor working conditions combined with intense stress cost Euler, then still only in his twenties, the sight of one eye. This is apparent in many portraits of Euler, including the one at the front of this chapter.

On the advice of Jean Le Rond d'Alembert, Euler was replaced by Joseph-Louis Lagrange as mathematician to the court of Frederick the Great, who later commented: 'To your care and recommendation am I indebted for having replaced a half-blind mathematician with a mathematician with both eyes, which will especially please the anatomical members of my Academy.' Euler returned to Russia where Catherine the Great welcomed back her 'mathematical cyclops'.

The loss of one eye was only a minor handicap – in fact Euler claimed that 'now I will have less distraction'. Forty years later, at the age of sixty, his situation worsened considerably, when a cataract in Euler's good eye meant he was destined to become completely blind. He was determined not to give in and began to practise writing with his fading eye closed in order to perfect his technique before the onset of darkness. Within weeks he was blind. The rehearsal paid off for a while, but a few months later Euler's script became illegible, whereupon his son Albert acted as his amanuensis.

Euler continued to produce mathematics for the next seventeen years and, if anything, he was more productive than ever. His immense intellect allowed him to juggle concepts without having to commit them to paper, and his phenomenal memory allowed him to use his own brain as a mental library. Colleagues suggested that the onset of blindness appeared to expand the horizons of his imagination. It is worth noting that Euler's computations of lunar positions were completed during his period of blindness. For the emperors of Europe this was the most prized of mathematical achievements, a problem that had confounded the greatest mathematicians in Europe, including Newton.

In 1776 an operation was performed to remove the cataract, and for a few days Euler's sight seemed to have been restored.

Then infection set in and Euler was plunged back into darkness. Undaunted he continued to work until, on 18 September 1783, he suffered a fatal stroke. In the words of the mathematician-philosopher the Marquis de Condorcet, 'Euler ceased to live and calculate.'

A Petty Pace

A century after Fermat's death there existed proofs for only two specific cases of the Last Theorem. Fermat had given mathematicians a head start by providing them with the proof that there were no solutions to the equation

$$x^4 + y^4 = z^4.$$

Euler had adapted the proof to show that there were no solutions to

$$x^3 + y^3 = z^3$$

After Euler's breakthrough it was still necessary to prove that there were no whole number solutions to an infinity of equations:

$$x^5 + y^5 = z^5,$$
$$x^6 + y^6 = z^6,$$
$$x^7 + y^7 = z^7,$$
$$x^8 + y^8 = z^8,$$
$$x^9 + y^9 = z^9,$$
$$\vdots$$

Although mathematicians were making embarrassingly slow progress, the situation was not quite as bad as it might seem at first sight. The proof for the case $n = 4$ also proves the cases $n = 8, 12, 16, 20, \dots$. The reason is that any number which can be written as an 8th (or a 12th, 16th, 20th, …) power can also be rewritten as a 4th power. For instance, the number 256 is equal to 2^8, but it is also equal to 4^4. Therefore any proof which works for the 4th power will also work for the 8th power and for any other power that is a multiple of 4. Using the same principle, Euler's proof for the case $n = 3$ automatically proves the cases $n = 6, 9, 12, 15, \dots$

Suddenly, the numbers are tumbling and Fermat looks vulnerable. The proof for the case $n = 3$ is particularly significant because the number 3 is an example of a *prime number*. As explained earlier, a prime number has the special property of not being the multiple of any whole number except for 1 and itself. Other prime numbers are 5, 7, 11, 13, …. All the remaining numbers are multiples of the primes, and are referred to as non-primes, or composite numbers.

Number theorists consider prime numbers to be the most important numbers of all because they are the atoms of mathematics. Prime numbers are the numerical building blocks because all other numbers can be created by multiplying combinations of the prime numbers. This seems to lead to a remarkable breakthrough. To prove Fermat's Last Theorem for all values of n, one merely has to prove it for the prime values of n. All other cases are merely multiples of the prime cases and would be proved implicitly.

Intuitively this enormously simplifies the problem, because you can ignore those equations which involve a value of n that is not a prime number. The number of equations remaining is now vastly reduced. For example, for the values of n up to 20, there are only six values which need to be proved:

$$x^5 + y^5 = z^5,$$
$$x^7 + y^7 = z^7,$$
$$x^{11} + y^{11} = z^{11},$$
$$x^{13} + y^{13} = z^{13},$$
$$x^{17} + y^{17} = z^{17},$$
$$x^{19} + y^{19} = z^{19}.$$

If one can prove Fermat's Last Theorem for just the prime values of n, then the theorem is proved for all values of n. If one considers all whole numbers, then it is obvious that there are infinitely many. If one considers just the prime numbers, which are only a small fraction of all the whole numbers, then surely the problem is much simpler?

Intuition would suggest that if you begin with an infinite quantity and then remove the bulk of it, then you would expect to be left with something finite. Unfortunately intuition is not the arbiter of truth in mathematics, but rather logic. In fact, it is possible to prove that the list of primes is never-ending. Therefore, despite being able to ignore the vast majority of equations relating to non-prime values of n, the remaining equations relating to prime values of n are still infinite in number.

The proof that there is an infinity of primes dates all the way back to Euclid, and is one of the classic arguments of mathematics. Initially Euclid assumes that there is a finite list of known prime numbers, and then shows that there must exist an infinite number of additions to this list. There are N prime numbers in Euclid's finite list, which are labelled $P_1, P_2, P_3, ..., P_N$. Euclid can then generate a new number Q_A such that

$$Q_A = (P_1 \times P_2 \times P_3 \times \cdots \times P_N) + 1.$$

This new number Q_A is either prime or not prime. If it is prime then we have succeeded in generating a new, bigger prime number, and therefore our original list of primes was not complete. On the other hand, if Q_A is not prime, then it must be perfectly divisible by a prime. This prime cannot be one of the known primes because dividing Q_A by any of the known primes will inevitably lead to a remainder of 1. Therefore there must be some new prime, which we can call P_{N+1}.

We have now arrived at the stage where either Q_A is a new prime or we have another new prime P_{N+1}. Either way we have added to our original list of primes. We can now repeat the process, including our new prime (P_{N+1} or Q_A) in our list, and generate some new number Q_B. Either this new number will be yet another new prime, or there will have to be some other new prime P_{N+2} that is not on our list of known primes. The upshot of the argument is that, however long our list of prime numbers, it is always possible to find a new one. Therefore the list of primes is never-ending and infinite.

But how can something which is undeniably smaller than an infinite quantity also be infinite? The German mathematician David Hilbert once said: 'The infinite! No other question has ever moved so profoundly the spirit of man; no other idea has so fruitfully stimulated his intellect; yet no other concept stands in greater need of clarification than that of the infinite.' To resolve the paradox of the infinite it is necessary to define what is meant by infinity. Georg Cantor, who worked alongside Hilbert, defined infinity as the size of the never-ending list of counting numbers (1, 2, 3, 4,...). Consequently anything which is comparable in size is equally infinite.

By this definition the number of even counting numbers, which would intuitively appear to be smaller, is also infinite. It is easy to demonstrate that the quantity of counting numbers and the quantity of even numbers are comparable because we can pair off each counting number with a corresponding even number:

$$1 \quad 2 \quad 3 \quad 4 \quad 5 \quad 6 \quad 7\cdots$$
$$\Downarrow \quad \Downarrow \quad \Downarrow \quad \Downarrow \quad \Downarrow \quad \Downarrow \quad \Downarrow\cdots$$
$$2 \quad 4 \quad 6 \quad 8 \quad 10 \quad 12 \quad 14\cdots$$

If every member of the counting numbers list can be matched up with a member of the even numbers list then the two lists must be the same size. This method of comparison leads to some surprising conclusions, including the fact that there are an infinite number of primes. Although Cantor was the first person to tackle infinity in a formal way, he was initially heavily criticised by the mathematical community for his radical definition. Towards the end of his career the attacks became increasingly personal and this resulted in Cantor suffering mental illness and severe depression. Eventually, after his death, his ideas became widely accepted as the only consistent, accurate and powerful definition of infinity. As a tribute Hilbert said: 'No one shall drive us from the paradise Cantor has created for us.'

Hilbert went on to create an example of infinity, known as *Hilbert's Hotel*, which clearly illustrates its strange qualities. This hypothetical hotel has the desirable attribute of having an infinite number of rooms. One day a new guest arrives and is disappointed to learn that, despite the hotel's infinite size, all the rooms are occupied. Hilbert, the clerk, thinks for a while and then reassures the new arrival that he will find an empty room. He asks all his current guests to move to the next room, so that the guest in room 1 moves to room 2, the guest in room 2 moves to room 3, and so on. Everybody who was in the hotel still has a room, which allows the new arrival to slip into the vacant room 1. This shows that infinity plus one equals infinity.

The following night Hilbert has to deal with a much greater problem. The hotel is still full when an infinitely large coach arrives with an infinite number of new guests. Hilbert remains unperturbed and rubs his hands at the thought of infinitely more hotel bills. He asks all his current guests to move to the room which is double the number of their current room. So the guest in room 1 moves to room 2, the guest in room 2 moves to room 4, and so on. Everybody who was in the hotel still has a room and yet an infinite number of rooms, all the odd ones, have been vacated for the new arrivals. This shows that double infinity is still infinity.

Hilbert's Hotel seems to suggest that all infinities are as large as each other, because various infinities seem to be able to squeeze into the same infinite hotel – the infinity of even numbers can be matched up and compared with the infinity of all counting numbers. However, some infinities are indeed bigger than others. For example, any attempt to pair every rational number with every irrational number ends in failure, and in fact it can be proved that the infinite set of irrational numbers is larger than the infinite set of rational numbers. Mathematicians have had to develop a whole system of nomenclature to deal with the varying scales of infinity and conjuring with these concepts is one of today's hottest topics.

Although the infinity of primes dashed hopes for an early proof of Fermat's Last Theorem, a countless supply of prime numbers does have more positive implications in other areas such as espionage and the evolution of insects. Before returning to the quest for a proof of Fermat's Last Theorem it is worth briefly investigating the uses and abuses of primes.

Prime number theory is one of the few areas of pure mathematics that has found a direct application in the real world, namely in cryptography. Cryptography involves scrambling secret messages so that they can only be unscrambled by the receiver and not by anybody else who might intercept them. The scrambling process requires the use of a secret key, and traditionally unscrambling the message simply requires the receiver to apply the key in reverse. With this procedure the key is the weakest link in the chain of security. First, the receiver and the sender must agree on the details of the key and the exchange of this information is a risky process. If the enemy can intercept the key being exchanged, then they can unscramble all subsequent messages. Second, the keys must be regularly changed in order to maintain security, and each time this happens there is a risk of the new key being intercepted.

The problem of the key revolves around the fact that applying it one way will scramble the message, and applying it in reverse unscrambles the message – unscrambling a message is almost as easy as scrambling it. However, experience tells us that there are many everyday situations when unscrambling is far harder than scrambling – it is relatively easy to scramble an egg, but to unscramble it is far harder.

In the 1970s Whitfield Diffie and Martin Hellman came up with the idea of looking for a mathematical process which was easy to perform in one direction but incredibly difficult to perform in the opposite direction. Such a process would provide a perfect key. For example, I could have my own two-part key, and publish the scrambling half of it in a public directory. Then anybody could send me scrambled messages, but only I would know the unscrambling half of the key. Although everyone would have knowledge of the scrambling part of the key, it bears no relation to the unscrambling part of the key.

In 1977 Ronald Rivest, Adi Shamir and Leonard Adleman, a team of mathematicians and computer scientists at the

Massachusetts Institute of Technology, realised that prime numbers were the ideal basis for an easy-scramble/hard-unscramble process. In order to make my own personal key I would take two huge prime numbers, each one containing up to 80 digits, and then multiply them together to achieve an even larger non-prime number. In order to scramble messages all that is required is knowledge of the large non-prime number, whereas to unscramble the message you would need to know the two original prime numbers which were multiplied together, known as the prime factors. I can now publish the large non-prime number, the scrambling half of the key, and keep the two prime factors, the unscrambling half of the key, to myself. Importantly, even though everybody knows the large non-prime number, they would have immense difficulty in working out the two prime factors.

Taking a simpler example, I could hand out the non-prime number 589, which would enable everyone to scramble messages to me. I would keep the two prime factors of 589 secret, so that only I could unscramble the messages. If others could work out the two prime factors then they too could unscramble my messages, but even with this small number it is not obvious what the two prime factors are. In this case it would only take a few minutes on a desktop computer to figure out that the prime factors are actually 31 and 19 (31 × 19 = 589), and so my key would not remain secure for very long.

However, in reality the non-prime number which I would publish would have over a hundred digits, which makes the task of finding its prime factors effectively impossible. Even if the world's most powerful computers were used to split this huge non-prime number (the scrambling key) into its two prime factors (the unscrambling key) it would take several years to achieve the answer. Therefore, to foil foreign spies, I merely have to change my key on an annual basis. Once a year I announce my new giant non-prime number, and anybody who wants to try and unscramble my messages would then have to start all over again trying to compute the two prime factors.

As well as finding a role in espionage, prime numbers also appear in the natural world. The periodical cicadas, most notably *Magicicada septendecim*, have the longest life-cycle of any insect. Their unique life-cycle begins underground, where the nymphs patiently suck the juice from the roots of trees. Then, after 17 years of waiting the adult cicadas emerge from the ground, swarm in vast numbers and temporarily swamp the landscape. Within a few weeks they mate, lay their eggs and die.

The question which puzzled biologists was, Why is the cicada's life-cycle so long? And is there any significance to the life-cycle being a prime number of years? Another species, *Magicicada tredecim*, swarms every 13 years, implying that life-cycles lasting a prime number of years offer some evolutionary advantage.

One theory suggests that the cicada has a parasite which also goes through a lengthy life-cycle and which the cicada is trying to avoid. If the parasite has a life-cycle of, say, 2 years then the cicada wants to avoid a life-cycle which is divisible by 2, otherwise the parasite and the cicada will regularly coincide. Similarly, if the parasite has a life-cycle of 3 years then the cicada wants to avoid a life-cycle which is divisible by 3, otherwise the parasite and the cicada will once again regularly coincide. Ultimately, to avoid meeting its parasite the cicadas' best strategy is to have a long life-cycle lasting a prime number of years. Because nothing will divide into 17, *Magicicada septendecim* will rarely meet its parasite. If the parasite has a 2-year life-cycle they will only meet every 34 years, and if it has a longer life-cycle, say 16 years, then they will only meet every 272 (16 × 17) years.

In order to fight back, the parasite only has two life-cycles which will increase the frequency of coincidences – the annual cycle and the same 17-year cycle as the cicada. However, the parasite is unlikely to survive reappearing 17 years in a row, because for the first 16 appearances there will be no cicadas for it to parasitise. On the other hand, in order to reach the 17-year life-cycle, the generations of parasites would first have to evolve through the 16-year life-cycle. This would mean at some stage of evolution the parasite and cicada would not coincide for 272 years! In either case the cicadas long prime life-cycle protects it.

This might explain why the alleged parasite has never been found! In the race to keep up with the cicada, the parasite probably kept extending its life-cycle until it hit the 16-year hurdle. Then it failed to coincide for 272 years, by which time the lack of coinciding with cicadas had driven it to extinction. The result is a cicada with a 17-year life cycle, which it no longer needs because its parasite no longer exists.

Monsieur Le Blanc

By the beginning of the nineteenth century, Fermat's Last Theorem had already established itself as the most notorious problem in number theory. Since Euler's breakthrough there had been no further progress, but a dramatic announcement by a young Frenchwoman was to reinvigorate the pursuit of Fermat's lost proof. Sophie Germain lived in an era of chauvinism and prejudice, and in order to conduct her research she was forced to assume a false identity, study in terrible conditions and work in intellectual isolation.

Over the centuries women have been discouraged from

Sophie Germain

studying mathematics, but despite the discrimination there have been several female mathematicians who fought against the establishment and indelibly forged their names in the annals of mathematics. The first woman known to have made an impact on the subject was Theano in the sixth century BC, who began as one of

Pythagoras' students before becoming one of his foremost disciples and eventually marrying him. Pythagoras is known as the 'feminist philosopher' because he actively encouraged women scholars, Theano being just one of the twenty-eight sisters in the Pythagorean Brotherhood.

In later centuries the likes of Socrates and Plato would continue to invite women into their schools, but it was not until the fourth century AD that a woman mathematician founded her own influential school. Hypatia, the daughter of a mathematics professor at the University of Alexandria, was famous for giving the most popular discourses in the known world and for being the greatest of problem-solvers. Mathematicians who had been stuck for months on a particular problem would write to her seeking a solution, and Hypatia rarely disappointed her admirers. She was obsessed by mathematics and the process of logical proof, and when asked why she never married she replied that she was wedded to the truth. Ultimately her devotion to the cause of rationalism caused her downfall, when Cyril, the patriarch of Alexandria, began to oppress philosophers, scientists and mathematicians, whom he called heretics. The historian Edward Gibbon provided a vivid account of what happened after Cyril had plotted against Hypatia and turned the masses against her:

On a fatal day, in the holy season of Lent, Hypatia was torn from her chariot, stripped naked, dragged to the church, and inhumanely butchered by the hands of Peter the Reader and a troop of savage and merciless fanatics; her flesh was scraped from her bones with sharp oyster-shells, and her quivering limbs were delivered to the flames.

Soon after the death of Hypatia mathematics entered a period of stagnation and it was not until after the Renaissance that another woman made her name as a mathematician. Maria Agnesi was born in Milan in 1718 and, like Hypatia, was the daughter of a mathematician. She was acknowledged to be one of the finest mathematicians in Europe, particularly famous for her treatises on the tangents to curves. In Italian, curves were called *versiera*, a word derived from the Latin *vertere*, 'to turn', but it was also an abbreviation for *avversiera*, or 'wife of the Devil'. A curve studied by Agnesi (*versiera Agnesi*) was mistranslated into English as the 'witch of Agnesi', and in time the mathematician herself was referred to by the same title.

Although mathematicians across Europe acknowledged Agnesi's ability, many academic institutions, in particular the French Academy, refused to give her a research post. Institutionalised discrimination against women continued right through to the twentieth century, when Emmy Noether, described by Einstein as 'the most significant creative mathematical genius thus far produced since the higher education of women began', was denied a lectureship at the University of Göttingen. The majority of the faculty argued: 'How can it be allowed that a woman become a *Privatdozent*? Having become a *Privatdozent*, she can then become a professor and a member of the University Senate …. What will our soldiers think when they return to the University and find that they are expected to learn at the feet of a woman?' Her friend and mentor David Hilbert replied: 'Meine Herren, I do not see that the sex of the candidate is an argument against her admission as a *Privatdozent*. After all, the Senate is not a bathhouse.'

Later her colleague Edmund Landau was asked whether Noether was indeed a great woman mathematician, to which he replied: 'I can testify that she is a great mathematician, but that she is a woman, I cannot swear.'

In addition to suffering discrimination Noether had much else in common with other women mathematicians through the centuries, such as the fact that she too was the daughter of a mathematics professor. Many mathematicians, of both genders, are from mathematical families, giving rise to light-hearted rumours of a mathematical gene, but in the case of women the percentage is particularly high. The probable explanation is that most women with potential were never exposed to the subject or encouraged to pursue it, whereas those born to professors could hardly avoid being immersed in the numbers. Furthermore, Noether, like Hypatia, Agnesi and most other women mathematicians, never married, largely because it was not socially acceptable for women to pursue such careers and there were few men who were prepared to wed brides with such controversial backgrounds. The great Russian mathematician Sonya Kovalevsky is an exception to this rule, inasmuch as she arranged a marriage of convenience to Vladimir Kovalevsky, a man who was agreeable to a platonic relationship. For both parties the marriage allowed them to escape their families and concentrate on their researches, and in Sonya's case travelling alone around Europe was much easier once she was a respectable married woman.

Of all the European countries France displayed the most chauvinistic attitude towards educated women, declaring that mathematics was unsuitable for women and beyond their mental capacity. Although the salons of Paris dominated the mathematical world for most of the eighteenth and nineteenth centuries, only one woman managed to escape the constraints of French society and establish herself as a great number theorist. Sophie Germain revolutionised the study of Fermat's Last Theorem and made a contribution greater than any of the men who had gone before her.

Sophie Germain was born on 1 April 1776, the daughter of a merchant, Ambroise-François Germain. Outside of her work, her life was to be dominated by the turmoils of the French Revolution – the year she discovered her love of numbers the Bastille was stormed, and her study of calculus was shadowed by the Reign of Terror. Although her father was financially successful, Sophie's family were not members of the aristocracy.

Although ladies of Germain's social background were not actively encouraged to study mathematics, they were expected to have sufficient knowledge of the subject in order to be able to discuss the topic should it arise during polite conversation. To this end a series of textbooks were written to help young women get to grips with the latest developments in mathematics and science. Francesco Algarotti was the author of *Sir Isaac Newton's Philosophy Explain'd for the Use of Ladies*. Because Algarotti believed that women were only interested in romance, he attempted to explain Newton's discoveries through the flirtatious dialogue between a Marquise and her interlocutor. For example, the interlocutor outlines the inverse square law of gravitational attraction, whereupon the Marquise gives her own interpretation on this fundamental law of physics: 'I cannot help thinking … that this proportion in the squares of the distances of places … is observed even in love. Thus after eight days' absence love becomes sixty-four times less than it was the first day.'

Not surprisingly this gallant genre of books was not responsible for inspiring Sophie Germain's interest in mathematics. The event that changed her life occurred one day when she was browsing in her father's library and chanced upon Jean-Etienne Montucla's book *History of Mathematics*. The chapter that caught her imagination was Montucla's essay on the life of Archimedes. His account of Archimedes' discoveries was undoubtedly interesting, but what particularly kindled her fascination was the story surrounding his death. Archimedes had spent his life at Syracuse, studying mathematics in relative tranquillity, but when he was in his late seventies the peace was shattered by the invading Roman army. Legend has it that during the invasion Archimedes was so engrossed in the study of a geometric figure in the sand that he failed to respond to the questioning of a Roman soldier. As a result he was speared to death.

Germain concluded that if somebody could be so consumed by a geometric problem that it could lead to their death, then mathematics must be the most captivating subject in the world. She immediately set about teaching herself the basics of number theory and calculus, and soon she was working late into the night, studying the works of Euler and Newton. This sudden interest in such an unfeminine subject worried her parents. A friend of the family, Count Guglielmo Libri-Carrucci dalla Sommaja, told how Sophie's father confiscated her candles and clothes and removed any heating in order to discourage her from studying. Only a few years later in Britain, the young mathematician Mary Somerville would also have her candles confiscated by her father who maintained that 'we must put a stop to this, or we shall have Mary in a strait-jacket one of these days'.

In Germain's case she responded by maintaining a secret cache of candles and wrapping herself in bed-clothes. Libri-Carrucci wrote that the winter nights were so cold that the ink froze in the inkwell but Sophie continued regardless. She was described by some people as shy and awkward, but she was also immensely determined and eventually her parents relented and gave Sophie their blessing. Germain never married and throughout her career her father funded her research. For many years Germain continued to study alone because there were no mathematicians in the family who could introduce her to the latest ideas and her tutors refused to take her seriously.

Then, in 1794, the Ecole Polytechnique opened in Paris. It was founded as an academy of excellence to train mathematicians and scientists for the nation. This would have been an ideal place for Germain to develop her mathematical skills except for the fact that it was an institution reserved only for men. Her natural shyness prevented her from confronting the academy's governing body, so instead she resorted to covertly studying at the Ecole by assuming the identity of a former student at the academy, Monsieur Antoine-August Le Blanc. The academy's administration was unaware that the real Monsieur Le Blanc had left Paris and continued to print lecture notes and problems for him. Germain managed to obtain what was intended for Le Blanc and each week she would submit answers to the problems under her new pseudonym. Everything was going to plan until a couple of months later when the supervisor of the course, Joseph-Louis Lagrange, could no longer ignore the brilliance of Monsieur Le Blanc's answer sheets. Not only were Monsieur Le Blanc's solutions marvellously ingenious, but they showed a remarkable transformation in a student who had previously been notorious for his abysmal calculations. Lagrange, who was one of the finest mathematicians of the nineteenth century, requested a meeting with the reformed student and Germain was forced to reveal her true identity. Lagrange was astonished and pleased to meet the young woman and became her mentor and friend. At last Sophie Germain had a teacher who could inspire her, and with whom she could be open about her skills and ambitions.

Germain grew in confidence and she moved from solving problems in her coursework to studying unexplored areas of mathematics. Most importantly she became interested in number theory and inevitably she came to hear of Fermat's Last Theorem. She worked on the problem for several years, eventually reaching the stage where she believed she had made an important breakthrough. She needed to discuss her ideas with a fellow number

theorist and decided that she would go straight to the top and con-
sult the greatest number theorist in the world, the German mathematician Carl Friedrich Gauss.

Gauss is acknowledged as being one of the most brilliant mathematicians who has ever lived. While E.T. Bell referred to Fermat
as the 'Prince of Amateurs', he called Gauss the 'Prince of
Mathematicians'. Germain had first encountered his work through
studying his masterpiece *Disquisitiones arithmeticae*, the most important and wide-ranging treatise since Euclid's *Elements*. Gauss's work
influenced every area of mathematics, but strangely enough he
never published anything on Fermat's Last Theorem. In one letter
he even displayed contempt for the problem. His friend the
German astronomer Heinrich Olbers had written to Gauss
encouraging him to compete for a prize which had been offered by
the Paris Academy for a solution to Fermat's challenge: 'It seems
to me, dear Gauss, that you should get busy about this.' Two weeks
later Gauss replied, 'I am very much obliged for your news concerning the Paris prize. But I confess that Fermat's Last Theorem
as an isolated proposition has very little interest for me, for I could
easily lay down a multitude of such propositions, which one could
neither prove nor disprove.' Gauss was entitled to his opinion, but
Fermat had clearly stated that a proof existed and even the subsequent failed attempts to find the proof had generated innovative
new techniques, such as proof by 'infinite descent' and the use of
imaginary numbers. Perhaps in the past Gauss had tried and failed
to make any impact on the problem, and his response to Olbers
was merely a case of intellectual sour grapes. Nonetheless, when he
received Germain's letters he was sufficiently impressed by her
breakthrough that he temporarily forgot his ambivalence towards
Fermat's Last Theorem.

Seventy-five years earlier Euler had published his proof for the case $n = 3$, and ever since mathematicians had been trying in vain to prove other individual cases. However, Germain adopted a new strategy and described to Gauss a so-called general approach to the problem. In other words, her immediate goal was not to prove one particular case, but to say something about many cases at once. In her letter to Gauss she outlined a calculation which focused on a particular type of prime number p such that $(2p + 1)$ is also prime. Germain's list of primes includes 5, because 11 $(2 \times 5 + 1)$ is also prime; but it does not include 13, because 27 $(2 \times 13 + 1)$ is not prime.

For values of n equal to these Germain primes, she used an elegant argument to show that there were probably no solutions to the equation $x^n + y^n = z^n$. By 'probably' Germain meant that it was unlikely that any solutions existed, because if there was a solution then either x, y or z would be a multiple of n, and this would put a very tight restriction on any solutions. Her colleagues examined her list of primes one by one trying to prove that x, y or z could not be a multiple of n, thereby showing that for that particular value of n there could be no solutions.

In 1825 her method claimed its first complete success thanks to Gustav Lejeune-Dirichlet and Adrien-Marie Legendre, two mathematicians a generation apart. Legendre was a man in his seventies who had lived through the political turmoil of the French Revolution. His failure to support the government candidate for the Institut National led to the stopping of his pension, and by the time he made his contribution to Fermat's Last Theorem he was destitute. On the other hand, Dirichlet was an ambitious young number theorist who had only just turned twenty. Both of them independently were able to prove that the case $n = 5$ has no solutions, but they based their proofs on, and owed their success to, Sophie Germain.

Fourteen years later the French made another breakthrough. Gabriel Lamé made some further ingenious additions to Germain's method and proved the case for the prime $n = 7$. Germain had shown numbers theorists how to destroy an entire section of prime cases and now it was up to the combined efforts of her colleagues to continue proving Fermat's Last Theorem one case at a time.

Germain's work on Fermat's Last Theorem was to be her greatest contribution to mathematics but initially she was not credited for her breakthrough. When Germain wrote to Gauss she was still in her twenties, and although she had gained a reputation in Paris she feared that the great man would not take her seriously because of her gender. In order to protect herself Germain resorted once again to her pseudonym, signing her letters as Monsieur Le Blanc.

Her fear and respect for Gauss is shown in one of her letters to him: 'Unfortunately, the depth of my intellect does not equal the voracity of my appetite, and I feel a kind of temerity in troubling a man of genius when I have no other claim to his attention than an admiration necessarily shared by all his readers.' Gauss, unaware of his correspondent's true identity, attempted to put Germain at ease and replied: 'I am delighted that arithmetic has found in you so able a friend.'

Germain's contribution may have been forever wrongly attributed to the mysterious Monsieur Le Blanc were it not for the Emperor Napoleon. In 1806 Napoleon was invading Prussia and the French army was storming through one German city after another. Germain feared that the fate that befell Archimedes might also take the life of her other great hero Gauss, so she sent a message to her friend General Joseph-Marie Pernety, who was in charge of the advancing forces. She asked him to guarantee

Gauss's safety, and as a result the general took special care of the German mathematician, explaining to him that he owed his life to Mademoiselle Germain. Gauss was grateful but surprised, for he had never heard of Sophie Germain.

The game was up. In Germain's next letter to Gauss she reluctantly revealed her true identity. Far from being angry at the deception, Gauss wrote back to her with delight:

But how to describe to you my admiration and astonishment at seeing my esteemed correspondent Monsieur Le Blanc metamorphose himself into this illustrious personage who gives such a brilliant example of what I would find it difficult to believe. A taste for the abstract sciences in general and above all the mysteries of numbers is excessively rare: one is not astonished at it: the enchanting charms of this sublime science reveal themselves only to those who have the courage to go deeply into it. But when a person of the sex which, according to our customs and prejudices, must encounter infinitely more difficulties than men to familiarise herself with these thorny researches, succeeds nevertheless in surmounting these obstacles and penetrating the most obscure parts of them, then without doubt she must have the noblest courage, quite extraordinary talents and superior genius. Indeed nothing could prove to me in so flattering and less equivocal manner that the attractions of this science, which has enriched my life with so many joys, are not chimerical, as the predilection with which you have honoured it.

Sophie Germain's correspondence with Carl Gauss inspired much of her work, but in 1808 the relationship ended abruptly. Gauss had been appointed professor of astronomy at the University of Göttingen, his interest shifted from number theory to more applied mathematics, and he no longer bothered to return Germain's letters. Without her mentor her confidence began to wane, and within a year she abandoned pure mathematics.

Although she made no further contributions to proving Fermat's Last Theorem, she did embark on an eventful career as a physicist, a discipline in which she would again excel only to be confronted by the prejudices of the establishment. Her most important contribution to the subject was 'Memoir on the vibrations of elastic plates', a brilliantly insightful paper which laid the foundations for the modern theory of elasticity. As a result of this research and her work on Fermat's Last Theorem she received a medal from the Institut de France, and became the first woman who was not a wife of a member to attend lectures at the Academy of Sciences. Then towards the end of her life she re-established her relationship with Carl Gauss, who convinced the University of Göttingen to award her an honorary degree. Tragically, before the university could bestow the honour upon her, Sophie Germain died of breast cancer.

All things considered she was probably the most profoundly intellectual woman that France has ever produced. And yet, strange as it may seem, when the state official came to make out the death certificate of this eminent associate and co-worker of the most illustrious members of the French Academy of Science, he designated her as a *rentière-annuitant* (a single woman with no profession) – not as a *mathématicienne*. Nor is this all. When the Eiffel Tower was erected, in which the engineers were obliged to give special attention to the elasticity of the materials used, there were inscribed on this lofty structure the names of seventy-two savants. But one will not find in this list the name of that daughter of genius, whose researches contributed so much towards establishing the theory of the elasticity of metals – Sophie Germain. Was she excluded from this list for the same reason that Agnesi was ineligible for membership in the French Academy – because she was a woman? It would seem so. If such, indeed, was the case, more is the shame for those who were responsible for such ingratitude towards one who had deserved so well of science, and who by her achievements had won an enviable place in the hall of fame.

H. J. Mozans, 1913

The Sealed Envelopes

After the breakthrough of Sophie Germain the French Academy of Sciences offered a series of prizes, including a gold medal and 3,000 Francs to the mathematician who could finally put to rest the mystery of Fermat's Last Theorem. As well as the prestige of proving Fermat's Last Theorem there was now an immensely valuable reward attached to the challenge. The salons of Paris were full of rumours as to who was adopting which strategy and how close they were to announcing a result. Then, on 1 March 1847, the Academy held its most dramatic meeting ever.

The proceedings describe how Gabriel Lamé, who had proved the case $n = 7$ some years earlier, took the podium in front of the most eminent mathematicians of the age and proclaimed that he was on the verge of proving Fermat's Last Theorem. He admitted that his proof was still incomplete, but he outlined his method and predicted with relish that he would in the coming weeks publish a complete proof in the Academy's journal.

The entire audience was stunned, but as soon as Lamé left the floor Augustin Louis Cauchy, another of Paris's finest mathematicians, asked for permission to speak. Cauchy announced to the Academy that he had been working along similar lines to Lamé, and that he too was about to publish a complete proof.

Both Cauchy and Lamé realised that time was of the essence. Whoever would be first to submit a complete proof would receive the most prestigious and valuable prize in mathematics. Although

Gabriel Lamé

Augustin Cauchy

neither of them had a complete proof, the two rivals were keen to somehow stake a claim and so just three weeks after they had made their announcements they deposited sealed envelopes at the Academy. This was a common practice at the time which enabled mathematicians to go on record without revealing the exact details of their work. If a dispute should later arise regarding the originality of ideas, then a sealed envelope would provide the evidence needed to establish priority.

The anticipation built up throughout April as Cauchy and Lamé published tantalising but vague details of their proof in the proceedings of the Academy. Although the entire mathematical community was desperate to see the proof completed, many of them secretly hoped that it would be Lamé and not Cauchy who would win the race. By all accounts Cauchy was a self-righteous creature, a religious bigot and extremely unpopular with his colleagues. He was only tolerated at the Academy because of his brilliance.

Then, on 24 May, an announcement was made which put an end to the speculation. It was neither Cauchy nor Lamé who addressed the Academy but rather Joseph Liouville. Liouville shocked the entire audience by reading out the contents of a letter from the German mathematician Ernst Kummer.

Kummer was a number theorist of the highest order, but for much of his career a fierce patriotism fired by a hatred of Napoleon deflected him from his true calling. When Kummer was an infant the French army invaded his home town of Sorau, bringing with them an epidemic of typhus. Kummer's father was the town physician and within weeks he was taken by the disease. Traumatised by the experience Kummer swore to do his utmost to defend his country from further attack, and as soon as he left university he applied his intellect to the problem of plotting the trajectories of

Ernst Kummer

cannon-balls. Ultimately he taught the laws of ballistics at Berlin's war college.

In parallel with his military career Kummer actively pursued pure mathematical research and had been fully aware of the ongoing saga at the French Academy. He had read through the proceedings and analysed the few details that Cauchy and Lamé had dared to reveal. To Kummer it was obvious that the two Frenchmen were heading towards the same logical dead end, and he outlined his reasons in the letter which he sent to Liouville.

According to Kummer the fundamental problem was that the proofs of both Cauchy and Lamé relied on using a property of numbers known as unique factorisation. Unique factorisation states that there is only one possible combination of primes which will multiply together to give any particular number. For instance, the only combination of primes which will build the number 18 is as follows:

$$18 = 2 \times 3 \times 3.$$

Similarly, the following numbers are uniquely factorised in the following ways:

$$35 = 5 \times 7,$$
$$180 = 2 \times 2 \times 3 \times 3 \times 5,$$
$$106{,}260 = 2 \times 2 \times 3 \times 5 \times 7 \times 11 \times 23.$$

Unique factorisation was discovered back in the fourth century BC by Euclid, who proved that it is true for all counting numbers and described the proof in Book IX of his *Elements*. The fact that unique factorisation is true for all counting numbers is a vital element in many other proofs and is nowadays called the *fundamental theorem of arithmetic*.

At first sight there should have been no reason why Cauchy and Lamé should not rely on unique factorisation, as had hundreds of mathematicians before them. Unfortunately both of their proofs involved imaginary numbers. Although unique factorisation is true for real numbers, Kummer pointed out that it might not necessarily hold true when imaginary numbers are introduced. According to him this was a fatal flaw.

For example, if we restrict ourselves to real numbers then the number 12 can only be factorised into $2 \times 2 \times 3$. However, if we allow imaginary numbers into our proof then 12 can also be factorised in the following way:

$$12 = (1 + \sqrt{-11}) \times (1 - \sqrt{-11}).$$

Here $(1 + \sqrt{-11})$ is a complex number, a combination of a real and an imaginary number. Although the process of multiplication is more convoluted than for ordinary numbers, the existence of complex numbers does lead to additional ways to factorise 12. Another way to factorise 12 is $(2 + \sqrt{-8}) \times (2 - \sqrt{-8})$. There is no longer a unique factorisation but rather a choice of factorisations.

This loss of unique factorisation severely damaged the proofs of Cauchy and Lamé, but it did not necessarily destroy them completely. The proofs were supposed to show that there were no solutions to the equation $x^n + y^n = z^n$, where n represents any number greater than 2. As discussed earlier in this chapter, the proof only had to work for the prime values of n. Kummer showed that by employing extra techniques it was possible to restore unique factorisation for various values of n. For example, the problem of unique factorisation could be circumvented for all prime numbers up to and including $n = 31$. However, the prime number $n = 37$ could not be dealt with so easily. Among the other primes less than 100, two others, $n = 59$ and 67, were also awkward cases. These so-called irregular primes, which are sprinkled throughout the remaining prime numbers, were now the stumbling block to a complete proof.

Kummer pointed out that there was no known mathematics which could tackle all these irregular primes in one fell swoop. However, he did believe that, by carefully tailoring techniques to each individual irregular prime, they could be dealt with one by one. Developing these customised techniques would be a slow and painful exercise, and worse still the number of irregular primes is still infinite. Disposing of them individually would occupy the world's community of mathematicians until the end of time.

Kummer's letter had a devastating effect on Lamé. With hindsight the assumption of unique factorisation was at best over-optimistic and at worst foolhardy. Lamé realised that had he been more open about his work he might have spotted the error sooner, and he wrote to his colleague Dirichlet in Berlin: 'If only you had been in Paris, or I had been in Berlin, all of this would not have happened.'

While Lamé felt humiliated, Cauchy refused to accept defeat. He felt that compared to Lamé's proof his own approach was less reliant on unique factorisation, and until Kummer's analysis had been fully checked there was the possibility that it was flawed. For several weeks he continued to publish articles on the subject, but by the end of the summer he too fell silent.

Kummer had demonstrated that a complete proof of Fermat's Last Theorem was beyond the current mathematical approaches. It was a brilliant piece of mathematical logic, but a massive blow to an entire generation of mathematicians who had hoped that they might solve the world's hardest mathematical problem.

The situation was summarised by Cauchy, who in 1857 wrote the Academy's closing report on their prize for Fermat's Last Theorem:

Report on the competition for the Grand Prize in mathematical sciences.
Already set in the competition for 1853 and prorogued to 1856.

Eleven memoirs have been presented to the secretary. But none has
solved the proposed question. Thus, after many times being put forward
for a prize, the question remains at the point where Monsieur Kummer
left it. However, the mathematical sciences should congratulate themselves for the works which were undertaken by the geometers, with their
desire to solve the question, specially by Monsieur Kummer; and the
Commissaries think that the Academy would make an honorable and useful decision if, by withdrawing the question from the competition, it would
adjugate the medal to Monsieur Kummer, for his beautiful researches on
the complex numbers composed of roots of unity and integers.

For over two centuries every attempt to rediscover the proof of
Fermat's Last Theorem had ended in failure. Throughout his
teenage years Andrew Wiles had studied the work of Euler,
Germain, Cauchy, Lamé and finally Kummer. He hoped he could
learn by their mistakes, but by the time he was an undergraduate
at the University of Oxford he confronted the same brick wall that
faced Kummer.

Some of Wiles's contemporaries were beginning to suspect that
the problem might be impossible. Perhaps Fermat had deceived
himself and therefore the reason why nobody had rediscovered
Fermat's proof was that no such proof existed. Despite this scepticism Wiles continued to search for a proof. He was inspired by the
knowledge that there had been several cases in the past of proofs
which had eventually been discovered only after centuries of effort.
And in some of those cases the flash of insight which solved the
problem did not rely on new mathematics; rather it was a proof
which could have been done long ago.

One example of a problem which evaded solution for decades is

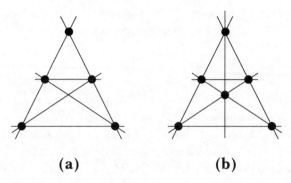

(a)   **(b)**

Figure 13. In these diagrams every dot is connected to every other dot by straight lines. Is it possible to construct a diagram such that every line has at least three dots on it?

the *dot conjecture*. The challenge involves a series of dots which are all connected to each other by straight lines, such as the dot diagrams shown in Figure 13. The conjecture claims that it is impossible to draw a dot diagram such that every line has at least three dots on it (excluding the diagram where all the dots are on the same line). Certainly by experimenting with a few diagrams this appears to be true. For example, Figure 13(a) has five dots connected by six lines. Four of the lines do not have three dots on them and so clearly this arrangement does not satisfy the requirement that all lines have three dots. By adding an extra dot and the associated line, as in Figure 13(b), the number of lines which do not have three dots is reduced to just three. However, trying to adapt the diagram further so that all the lines have three dots appears to be impossible. Of course, this does not prove that no such diagram exists.

Generations of mathematicians tried and failed to find a proof of the apparently straightforward dot conjecture. What made the conjecture even more infuriating was that when a proof was eventually discovered, it involved only a minimal amount of mathematical knowledge mixed with a little extra cunning. The proof is outlined in Appendix 6.

There was a possibility that all the techniques required to prove Fermat's Last Theorem were available, and that the only missing ingredient was ingenuity. Wiles was not prepared to give up: finding a proof of the Last Theorem had turned from being a childhood fascination in to a fully fledged obsession. Having learnt all there was to learn about the mathematics of the nineteenth century, Wiles decided to arm himself with techniques of the twentieth century.

Paul Wolfskehl

4

Into Abstraction

Proof is an idol before which the mathematician tortures himself.

Sir Arthur Eddington

Following the work of Ernst Kummer, hopes of finding a proof for the Last Theorem seemed fainter than ever. Furthermore mathematics was beginning to move into different areas of study and there was a risk that the new generation of mathematicians would ignore what seemed an impossible dead-end problem. By the beginning of the twentieth century the problem still held a special place in the hearts of number theorists, but they treated Fermat's Last Theorem in the same way that chemists treated alchemy. Both were foolish romantic dreams from a past age.

Then in 1908 Paul Wolfskehl, a German industrialist from Darmstadt, gave the problem a new lease of life. The Wolfskehl family were famous for their wealth and their patronage of the arts and sciences, and Paul was no exception. He had studied mathematics at university and, although he devoted most of his life to building the family's business empire, he maintained contact with professional mathematicians and continued to dabble in number theory. In particular Wolfskehl refused to give up on Fermat's Last Theorem.

Wolfskehl was by no means a gifted mathematician and he was

not destined to make a major contribution to finding a proof of the Last Theorem. Nonetheless, thanks to a curious chain of events, he was to become forever associated with Fermat's problem, and would inspire thousands of others to take up the challenge.

The story begins with Wolfskehl's obsession with a beautiful woman, whose identity has never been established. Depressingly for Wolfskehl the mysterious woman rejected him and he was left in such a state of utter despair that he decided to commit suicide. He was a passionate man, but not impetuous, and he planned his death with meticulous detail. He set a date for his suicide and would shoot himself through the head at the stroke of midnight. In the days that remained he settled all his outstanding business affairs, and on the final day he wrote his will and composed letters to all his close friends and family.

Wolfskehl had been so efficient that everything was completed slightly ahead of his midnight deadline, so to while away the hours he went to the library and began browsing through the mathematical publications. It was not long before he found himself staring at Kummer's classic paper explaining the failure of Cauchy and Lamé. It was one of the great calculations of the age and suitable reading for the final moments of a suicidal mathematician. Wolfskehl worked through the calculation line by line. Suddenly he was startled at what appeared to be a gap in the logic – Kummer had made an assumption and failed to justify a step in his argument. Wolfskehl wondered whether he had uncovered a serious flaw or whether Kummer's assumption was justified. If the former were true, then there was a chance that proving Fermat's Last Theorem might be a good deal easier than many had presumed.

He sat down, explored the inadequate segment of the proof, and became engrossed in developing a mini-proof which would either consolidate Kummer's work or prove that his assumption was wrong, in which case all Kummer's work would be invalidated. By dawn his work was complete. The bad news, as far as mathematics was concerned, was that Kummer's proof had been remedied and the Last Theorem remained in the realm of the unattainable. The good news was that the appointed time of the suicide had passed, and Wolfskehl was so proud that he had discovered and corrected a gap in the work of the great Ernst Kummer that his despair and sorrow evaporated. Mathematics had renewed his desire for life.

Wolfskehl tore up his farewell letters and rewrote his will in the light of what had happened that night. Upon his death in 1908 the new will was read out, and the Wolfskehl family were shocked to discover that Paul had bequeathed a large proportion of his fortune as a prize to be awarded to whomsoever could prove Fermat's Last Theorem. The reward of 100,000 Marks, worth over £1,000,000 in today's money, was his way of repaying a debt to the conundrum that had saved his life.

The money was put into the charge of the *Königliche Gesellschaft der Wissenschaften* of Göttingen, which officially announced the competition for the Wolfskehl Prize that same year:

By the power conferred on us, by Dr. Paul Wolfskehl, deceased in Darmstadt, we hereby fund a prize of one hundred thousand Marks, to be given to the person who will be the first to prove the great theorem of Fermat.

The following rules will be followed:

(1) The *Königliche Gesellschaft der Wissenschaften* in Göttingen will have absolute freedom to decide upon whom the prize should be conferred. It will refuse to accept any manuscript written with the sole aim of entering the competition to obtain the Prize. It will only take into consideration those mathematical memoirs which have appeared in the form of a

monograph in the periodicals, or which are for sale in the bookshops. The Society asks the authors of such memoirs to send at least five printed exemplars.

(2) Works which are published in a language which is not understood by the scholarly specialists chosen for the jury will be excluded from the competition. The authors of such works will be allowed to replace them by translations, of guaranteed faithfulness.

(3) The Society declines responsibility for the examination of works not brought to its attention, as well as for the errors which might result from the fact that the author of a work, or part of a work, are unknown to the Society.

(4) The Society retains the right of decision in the case where various persons would have dealt with the solution of the problem, or for the case where the solution is the result of the combined efforts of several scholars, in particular concerning the partition of the Prize.

(5) The award of the Prize by the Society will take place not earlier than two years after the publication of the memoir to be crowned. The interval of time is intended to allow German and foreign mathematicians to voice their opinion about the validity of the solution published.

(6) As soon as the Prize is conferred by the Society, the laureate will be informed by the secretary, in the name of the Society; the result will be published wherever the Prize has been announced during the preceding year. The assignment of the Prize by the Society is not to be the subject of any further discussion.

(7) The payment of the Prize will be made to the laureate, in the next three months after the award, by the Royal Cashier of Göttingen University, or, at the receiver's own risk, at any other place he may have designated.

(8) The capital may be delivered against receipt, at the Society's will, either in cash, or by the transfer of financial values. The payment of the Prize will be considered as accomplished by the transmission of these financial values, even though their total value at the day's end may not attain 100,000 Marks.

(9) If the Prize is not awarded by 13 September 2007, no ulterior claim will be accepted.

The competition for the Wolfskehl Prize is open, as of today, under the above conditions.

Göttingen, 27 June 1908
Die Königliche Gesellschaft der Wissenschaften

It is worth noting that although the Committee would give 100,000 Marks to the first mathematician to prove that Fermat's Last Theorem is true, they would not award a single pfennig to anybody who might prove that it is false.

The Wolfskehl Prize was announced in all the mathematical journals and news of the competition rapidly spread across Europe. Despite the publicity campaign and the added incentive of an enormous prize the Wolfskehl Committee failed to arouse a great deal of interest among serious mathematicians. The majority of professional mathematicians viewed Fermat's Last Theorem as a lost cause and decided that they could not afford to waste their careers working on a fool's errand. However, the prize did succeed in introducing the problem to a whole new audience, a hoard of eager minds who were willing to apply themselves to the ultimate riddle and approach it from a path of complete innocence.

The Era of Puzzles, Riddles and Enigmas

Ever since the Greeks, mathematicians have sought to spice up their textbooks by rephrasing proofs and theorems in the form of solutions to number puzzles. During the latter half of the nineteenth century this playful approach to the subject found its way into the popular press, and number puzzles were to be found alongside crosswords and anagrams. In due course there was a growing audience for mathematical conundrums, as amateurs contemplated everything from the most trivial riddles to profound mathematical problems, including Fermat's Last Theorem.

Perhaps the most prolific creator of riddles was Henry Dudeney, who wrote for dozens of newspapers and magazines, including the *Strand*, *Cassell's*, the *Queen*, *Tit-Bits*, the *Weekly Dispatch* and *Blighty*. Another of the great puzzlers of the Victorian Age was the Reverend Charles Dodgson, lecturer in mathematics at Christ Church, Oxford, and better known as the author Lewis Carroll. Dodgson devoted several years to compiling a giant compendium of puzzles entitled *Curiosa Mathematica*, and although the series was not completed he did write several volumes, including *Pillow Problems*.

The greatest riddler of them all was the American prodigy Sam Loyd (1841–1911), who as a teenager was making a healthy profit by creating new puzzles and reinventing old ones. He recalls in *Sam Loyd and his Puzzles: An Autobiographical Review* that some of his early puzzles were created for the circus owner and trickster P.T. Barnum:

Many years ago, when Barnum's Circus was of a truth 'the greatest show on earth', the famous showman got me to prepare for him a series of prize puzzles for advertising purposes. They became widely known as the 'Questions of the Sphinx', on account of the large prizes offered to anyone who could master them.

Strangely this autobiography was written in 1928, seventeen years after Loyd's death. Loyd passed his cunning on to his son, also called Sam, who was the real author of the book, knowing full well that anybody buying it would mistakenly assume that it had been written by the more famous Sam Loyd Senior.

Figure 14. A cartoon reflecting the mania caused by Sam Loyd's '14–15' puzzle.

Loyd's most famous creation was the Victorian equivalent of the Rubik's Cube, the '14–15' puzzle, which is still found in toyshops today. Fifteen tiles numbered 1 to 15 are arranged in a 4 × 4 grid, and the aim is to slide the tiles and rearrange them into the correct order. Loyd's '14–15' puzzle was sold in the arrangement shown in Figure 14, and he offered a significant reward to whoever could complete the puzzle by swapping the '14' and '15' into their proper positions via any series of tile slides. Loyd's son wrote about the fuss generated by this tangible but essentially mathematical puzzle:

A prize of $1,000, offered for the first correct solution to the problem, has never been claimed, although there are thousands of persons who say they performed the required feat. People became infatuated with the puzzle and ludicrous tales are told of shopkeepers who neglected to open their stores; of a distinguished clergyman who stood under a street lamp all through a wintry night trying to recall the way he had performed the feat. The mysterious feature of the puzzle is that none seem to be able to remember the sequence of moves whereby they feel sure they succeeded in solving the puzzle. Pilots are said to have wrecked their ships, and engineers rushed their trains past stations. A famous Baltimore editor tells how he went for his noon lunch and was discovered by his frantic staff long past midnight pushing little pieces of pie around on a plate!

Loyd was always confident that he would never have to pay out the $1,000 because he knew that it is impossible to swap just two pieces without destroying the order elsewhere in the puzzle. In the same way that a mathematician can prove that a particular equation has no solutions, Loyd could prove that his '14–15' puzzle is insoluble.

Loyd's proof began by defining a quantity which measured how disordered a puzzle is, the disorder parameter D_p. The disorder parameter for any given arrangement is the number of tile pairs which are in the wrong order, so for the correct puzzle, as shown in Figure 15(a), $D_p = 0$, because no tiles are in the wrong order.

By starting with the ordered puzzle and then sliding the tiles around, it is relatively easy to get to the arrangement shown in Figure 15(b). The tiles are in the correct order until we reach tiles 12 and 11. Obviously the 11 tile should come before the 12 tile and so this pair of tiles is in the wrong order. The complete list of tile pairs which are in the wrong order is as follows: (12,11), (15,13), (15,14), (15,11), (13,11) and (14,11). With six tile pairs in the wrong order in this arrangment, $D_p = 6$. (Note that tile 10 and tile 12 are

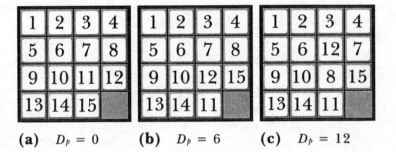

(a) $D_p = 0$ (b) $D_p = 6$ (c) $D_p = 12$

Figure 15. By sliding the tiles it is possible to create various disordered arrangements. For each arrangement it is possible to measure the amount of disorder, via the disorder parameter, D_p.

next to each other, which is clearly incorrect, but they are not in the wrong order. Therefore this tile pair does not contribute to the disorder parameter.)

After a bit more sliding we get to the arrangement in Figure 15(c). If you compile a list of tile pairs in the wrong order then you will discover that $D_p = 12$. The important point to notice is that in all these cases, (a), (b) and (c), the value of the disorder parameter is an even number (0, 6 and 12). In fact, if you begin with the correct arrangement and proceed to rearrange it, then this statement is always true. As long as the empty square ends up in the bottom right-hand corner, any amount of tile sliding will always result in an even value for D_p. The even value for the disorder parameter is an integral property of any arrangement derived from the original correct arrangement. In mathematics a property which always holds true no matter what is done to the object is called an *invariant*.

However, if you examine the arrangement which was being sold by Loyd, in which the 14 and 15 were swapped, then the value of the disorder parameter is one, $D_p = 1$, i.e. the only pair of tiles out of order are the 14 and 15. For Loyd's arrangement the disorder parameter has an odd value! Yet we know that any arrangement derived from the correct arrangement has an even value for the disorder parameter. The conclusion is that Loyd's arrangement cannot be derived from the correct arrangement, and conversely it is impossible to get from Loyd's arrangement back to the correct one – Loyd's $1,000 was safe.

Loyd's puzzle and the disorder parameter demonstrate the power of an invariant. Invariants provide mathematicians with an important strategy to prove that it is impossible to transform one object into another. For example, an area of current excitement concerns the study of knots, and naturally knot theorists are interested in trying to prove whether or not one knot can be transformed into another by twisting and looping but without cutting. In order to answer this question they attempt to find a property of the first knot which cannot be destroyed no matter how much twisting and looping occurs – a knot invariant. They then calculate the same property for the second knot. If the values are different then the conclusion is that it must be impossible to get from the first knot to the second.

Until this technique was invented in the 1920s by Kurt Reidemeister it was impossible to prove that one knot could not be transformed into any other knot. In other words before knot invariants were discovered it was impossible to prove that a granny knot is fundamentally different from a reef knot, an overhand knot or even a simple loop with no knot at all. The concept of an invariant property is central to many other mathematical proofs and, as we shall see in Chapter 5, it would be crucial in bringing Fermat's Last Theorem back into the mainstream of mathematics.

By the turn of the century, thanks to the likes of Sam Loyd and his '14–15' puzzle, there were millions of amateur problem-solvers throughout Europe and America eagerly looking for new challenges. Once news of Wolfskehl's legacy filtered down to these budding mathematicians, Fermat's Last Theorem was once again the world's most famous problem. The Last Theorem was infinitely more complex than even the hardest of Loyd's puzzles, but the prize was vastly greater. Amateurs dreamed that they might be able to find a relatively simple trick which had eluded the great professors of the past. The keen twentieth-century amateur was to a large extent on a par with Pierre de Fermat when it came to knowledge of mathematical techniques. The challenge was to match the creativity with which Fermat used his techniques.

Within a few weeks of announcing the Wolfskehl Prize an avalanche of entries poured into the University of Göttingen. Not surprisingly all the proofs were fallacious. Although each entrant was convinced that they had solved this centuries-old problem they had all made subtle, and sometimes not so subtle, errors in their logic. The art of number theory is so abstract that it is frighteningly easy to wander off the path of logic and be completely unaware that one has strayed into absurdity. Appendix 7 shows the sort of classic error which can easily be overlooked by an enthusiastic amateur.

Regardless of who had sent in a particular proof, every single one of them had to be scrupulously checked just in case an unknown amateur had stumbled upon the most sought after proof in mathematics. The head of the mathematics department at Göttingen between 1909 and 1934 was Professor Edmund Landau and it was his responsibility to examine the entries for the Wolfskehl Prize. Landau found that his research was being continually interrupted by having to deal with the dozens of confused proofs which arrived on his desk each month. To cope with the

situation he invented a neat method of off-loading the work. The
professor printed hundreds of cards which read:

> Dear ....,
>
> Thank you for your manuscript on the proof of
> Fermat's Last Theorem.
> The first mistake is on :
> Page .... Line ....
> This invalidates the proof.
>
> Professor E. M. Landau

Landau would then hand each new entry, along with a printed
card, to one of his students and ask them to fill in the blanks.

The entries continued unabated for years, even following the
dramatic devaluation of the Wolfskehl Prize – the result of the
hyperinflation which followed the First World War. There are
rumours which say that anyone winning the competition today
would hardly be able to purchase a cup of coffee with the prize
money, but these claims are somewhat exaggerated. A letter written by Dr F. Schlichting, who was responsible for dealing with
entries during the 1970s, explains that the prize was then still
worth over 10,000 Marks. The letter, written to Paulo Ribenboim
and published in his book *13 Lectures on Fermat's Last Theorem*, gives
a unique insight into the work of the Wolfskehl committee:

Dear Sir,

There is no count of the total number of 'solutions' submitted so
far. In the first year (1907–1908) 621 solutions were registered in the

files of the *Akademie*, and today they have stored about 3 metres of correspondence concerning the Fermat problem. In recent decades it was handled in the following way: the secretary of the *Akademie* divides the arriving manuscripts into:

(1) complete nonsense, which is sent back immediately,

(2) material which looks like mathematics.

The second part is given to the mathematical department, and there the work of reading, finding mistakes and answering is delegated to one of the scientific assistants (at German universities these are graduated individuals working for their Ph.D.) – at the moment I am the victim. There are about 3 or 4 letters to answer each month, and this includes a lot of funny and curious material, e.g. like the one sending the first half of his solution and promising the second if we would pay 1,000 DM in advance; or another one, who promised me 1% of his profits from publications, radio and TV interviews after he got famous, if only I would support him now; if not, he threatened to send it to a Russian mathematics department to deprive us of the glory of discovering him. From time to time someone appears in Göttingen and insists on personal discussion.

Nearly all 'solutions' are written on a very elementary level (using the notions of high school mathematics and perhaps some undigested papers in number theory), but can nevertheless be very complicated to understand. Socially, the senders are often persons with a technical education but a failed career who try to find success with a proof of the Fermat problem. I gave some of the manuscripts to physicians who diagnosed heavy schizophrenia.

One condition of Wolfskehl's last will was that the *Akademie* had to publish the announcement of the prize yearly in the main mathematical periodicals. But already after the first years the periodicals refused to print the announcement, because they were overflowed by letters and crazy manuscripts.

I hope that this information is of interest to you.

Yours sincerely,
F. Schlichting

As Dr Schlichting mentions, competitors did not restrict themselves to sending their 'solutions' to the *Akademie*. Every mathematics department in the world probably has its cupboard of purported proofs from amateurs. While most institutions ignore these amateur proofs, other recipients have dealt with them in more imaginative ways. The mathematical writer Martin Gardner recalls a friend who would send back a note explaining that he was not competent to examine the proof. Instead he would provide them with the name and address of an expert in the field who could help – that is to say, the details of the last amateur to send him a proof. Another of his friends would write: 'I have a remarkable refutation of your attempted proof, but unfortunately this page is not large enough to contain it.'

Although amateur mathematicians around the world have spent this century trying and failing to prove Fermat's Last Theorem and win the Wolfskehl Prize, the professionals have continued largely to ignore the problem. Instead of building on the work of Kummer and the other nineteenth-century number theorists, mathematicians began to examine the foundations of their subject in order to address some of the most fundamental questions about numbers. Some of the greatest figures of the twentieth century, including Bertrand Russell, David Hilbert and Kurt Gödel, tried to understand the most profound properties of numbers in order to grasp their true meaning and to discover what questions number theory can and, more importantly, cannot answer. Their work would shake the foundations of mathematics and ultimately have repercussions for Fermat's Last Theorem.

The Foundations of Knowledge

For hundreds of years mathematicians had been busy using logical proof to build from the known into the unknown. Progress had been phenomenal, with each new generation of mathematicians expanding on their grand structure and creating new concepts of number and geometry. However, towards the end of the nineteenth century, instead of looking forward, mathematical logicians began to look back to the foundations of mathematics upon which everything else was built. They wanted to verify the fundamentals of mathematics and rigorously rebuild everything from first principles, in order to reassure themselves that those first principles were reliable.

Mathematicians are notorious for being sticklers when it comes to requiring absolute proof before accepting any statement. Their reputation is clearly expressed in a story told by Ian Stewart in *Concepts of Modern Mathematics*:

An astronomer, a physicist, and a mathematician (it is said) were holidaying in Scotland. Glancing from a train window, they observed a black sheep in the middle of a field. 'How interesting,' observed the astronomer, 'all Scottish sheep are black!' To which the physicist responded, 'No, no! *Some* Scottish sheep are black!' The mathematician gazed heavenward in supplication, and then intoned, 'In Scotland there exists at least one field, containing at least one sheep, *at least one side of which is black.*'

Even more rigorous than the ordinary mathematician is the mathematician who specialises in the study of mathematical logic. Mathematical logicians began to question ideas which other mathematicians had taken for granted for centuries. For example, the law of trichotomy states that every number is either negative, positive or zero. This seems to be obvious and mathematicians had tacitly assumed it to be true, but nobody had ever bothered to prove that this really was the case. Logicians realised that, until the law of trichotomy had been proved true, then it might be false, and if that turned out to be the case then an entire edifice of knowledge, everything that relied on the law, would collapse. Fortunately for mathematics, at the end of the last century the law of trichotomy was proved to be true.

Ever since the ancient Greeks, mathematics had been accumulating more and more theorems and truths, and although most of them had been rigorously proved mathematicians were concerned that some of them, such as the law of trichotomy, had crept in without being properly examined. Some ideas had become part of the folklore and yet nobody was quite sure how they had been originally proved, if indeed they ever had been, so logicians decided to prove every theorem from first principles. However, every truth had to be deduced from other truths. Those truths, in turn, first had to be proved from even more fundamental truths, and so on. Eventually the logicians found themselves dealing with a few essential statements which were so fundamental that they themselves could not be proved. These fundamental assumptions are the axioms of mathematics.

One example of the axioms is the *commutative law of addition*, which simply states that, for any numbers m and n,

$$m + n = n + m.$$

This and the handful of other axioms are taken to be self-evident, and can easily be tested by applying them to particular numbers. So far the axioms have passed every test and have been accepted as being the bedrock of mathematics. The challenge for the logicians was to rebuild all of mathematics from these axioms.

Appendix 8 defines the set of arithmetic axioms and gives an idea of how logicians set about building the rest of mathematics.

A legion of logicians participated in the slow and painful process of rebuilding the immensely complex body of mathematical knowledge using only a minimal number of axioms. The idea was to consolidate what mathematicians thought they already knew by employing only the most rigorous standards of logic. The German mathematician Hermann Weyl summarised the mood of the time: 'Logic is the hygiene the mathematician practises to keep his ideas healthy and strong.' In addition to cleansing what was known, the hope was that this fundamentalist approach would also throw light on as yet unsolved problems, including Fermat's Last Theorem.

The programme was headed by the most eminent figure of the age, David Hilbert. Hilbert believed that everything in mathematics could and should be proved from the basic axioms. The result of this would be to demonstrate conclusively the two most important elements of the mathematical system. First, mathematics should, at least in theory, be able to answer every single question – this is the same ethos of completeness which had in the past demanded the invention of new numbers like the negatives and the imaginaries. Second, mathematics should be free of inconsistencies – that is to say, having shown that a statement is true by one method, it should not be possible to show that the same statement is false via another method. Hilbert was convinced that, by assuming just a few axioms, it would be possible to answer any imaginable mathematical question without fear of contradiction.

On 8 August 1900 Hilbert gave a historic talk at the International Congress of Mathematicians in Paris. Hilbert posed twenty-three unsolved problems in mathematics which he believed were of the most immediate importance. Some of the problems related to more general areas of mathematics, but most of them concentrated on the logical foundations of the subject. These problems were intended to focus the attention of the mathematical world and provide a programme of research. Hilbert wanted to galvanise the community into helping him realise his vision of a mathematical system free of doubt and inconsistency – an ambition he had inscribed on his tombstone:

> Wir müssen wissen,
> Wir werden wissen.
>
> We must know,
> We will know.

Although sometimes a bitter rival of Hilbert, Gottlob Frege was one of the leading lights in the so-called Hilbert programme. For over a decade Frege devoted himself to deriving hundreds of complicated theorems from the simple axioms, and his successes led him to believe that he was well on the way to completing a significant chunk of Hilbert's dream. One of Frege's key breakthroughs was to create the very definition of a number. For example, what do we actually mean by the number 3? It turns out that to define 3, Frege first had to define 'threeness'.

'Threeness' is the abstract quality which belongs to collections or sets of objects containing three objects. For instance 'threeness' could be used to describe the collection of blind mice in the popular nursery rhyme, or 'threeness' is equally appropriate for describing the set of sides of a triangle. Frege noticed that there were numerous sets which exhibited 'threeness' and used the idea of sets to define '3' itself. He created a new set and placed inside it all the sets exhibiting 'threeness' and called this new set of sets '3'. Therefore, a set has three members if and only if it is inside the set '3'.

This might appear to be an over-complex definition for a con-

David Hilbert

cept we use every day, but Frege's description of '3' is rigorous and indisputable and wholly necessary for Hilbert's uncompromising programme.

In 1902 Frege's ordeal seemed to be coming to an end as he prepared to publish *Grundgesetze der Arithmetik* (Fundamental Laws of Arithmetic) – a gigantic and authoritative two-volume work intended to establish a new standard of certainty within mathematics. At the same time the English logician Bertrand Russell, who was also contributing to Hilbert's great project, was making a devastating discovery. Despite following the rigorous protocol of Hilbert, he had come up against an inconsistency. Russell recalled his own reaction to the dreaded realisation that mathematics might be inherently contradictory:

At first I supposed that I should be able to overcome the contradiction quite easily, and that probably there was some trivial error in the reasoning. Gradually, however, it became clear that this was not the case . . . Throughout the latter half of 1901 I supposed the solution would be easy, but by the end of that time I had concluded that it was a big job . . . I made a practice of wandering about the common every night from eleven until one, by which time I came to know the three different noises made by nightjars. (Most people only know one.) I was trying hard to solve the contradiction. Every morning I would sit down before a blank sheet of paper. Throughout the day, with a brief interval for lunch, I would stare at the blank sheet. Often when evening came it was still empty.

There was no escaping from the contradiction. Russell's work would cause immense damage to the dream of a mathematical system free of doubt, inconsistency and paradox. He wrote to Frege, whose manuscript was already at the printers. The letter made Frege's life's work effectively worthless, but despite the mortal blow he published his *magnum opus* regardless and merely added

Bertrand Russell

a postscript to the second volume: 'A scientist can hardly meet with anything more undesirable than to have the foundation give way just as the work is finished. In this position I was put by a letter from Mr Bertrand Russell as the work was nearly through the press.'

Ironically Russell's contradiction grew out of Frege's much loved sets, or collections. Many years later, in his book *My Philosophical Development*, he recalled the thoughts that sparked his questioning of Frege's work: 'It seemed to me that a class sometimes is, and sometimes is not, a member of itself. The class of teaspoons, for example, is not another teaspoon, but the class of things that are not teaspoons is one of the things that are not teaspoons.' It was this curious and apparently innocuous observation that led to the catastrophic paradox.

Russell's paradox is often explained using the tale of the meticulous librarian. One day, while wandering between the shelves, the librarian discovers a collection of catalogues. There are separate catalogues for novels, reference, poetry, and so on. The librarian notices that some of the catalogues list themselves, while others do not.

In order to simplify the system the librarian makes two more catalogues, one of which lists all the catalogues which do list themselves and, more interestingly, one which lists all the catalogues which do not list themselves. Upon completing the task the librarian has a problem: should the catalogue which lists all the catalogues which do not list themselves, be listed in itself? If it is listed, then by definition it should not be listed. However, if it is not listed, then by definition it should be listed. The librarian is in a no-win situation.

The catalogues are very similar to the sets or classes which Frege used as the fundamental definition of numbers. Therefore the inconsistency which plagues the librarian will also cause problems in the supposedly logical structure of mathematics. Mathematics cannot tolerate inconsistencies, paradoxes or contradictions. For example, the powerful tool of proof by contradiction relies on a mathematics free of paradox. Proof by contradiction states that if an assumption leads to absurdity then the assumption must be false, but according to Russell even the axioms can lead to absurdity. Therefore proof by contradiction could show an axiom to be false, and yet axioms are the foundations of mathematics and acknowledged to be true.

Many intellectuals questioned Russell's work, claiming that mathematics was an obviously successful and unflawed pursuit. He responded by explaining the significance of his work in the following way:

'But,' you might say, 'none of this shakes my belief that 2 and 2 are 4.' You are quite right, except in marginal cases – and it is only in marginal cases that you are doubtful whether a certain animal is a dog or a certain length is less than a metre. Two must be two of something, and the proposition '2 and 2 are 4' is useless unless it can be applied. Two dogs and two dogs are certainly four dogs, but cases arise in which you are doubtful whether two of them are dogs. 'Well, at any rate there are four animals,' you might say. But there are microorganisms concerning which it is doubtful whether they are animals or plants. 'Well, then living organisms,' you say. But there are things of which it is doubtful whether they are living or not. You will be driven into say: 'Two entities and two entities are four entities.' When you have told me what you mean by 'entity', we will resume the argument.

Russell's work shook the foundations of mathematics and threw the study of mathematical logic into a state of chaos. The logicians were aware that a paradox lurking in the foundations of mathematics could sooner or later rear its illogical head and cause profound problems. Along with Hilbert and the other logicians, Russell set about trying to remedy the situation and restore sanity to mathematics.

This inconsistency was a direct consequence of working with the axioms of mathematics, which until this point had been assumed to be self-evident and sufficient to define the rest of mathematics. One approach was to create an additional axiom which forbade any class from being a member of itself. This would prevent Russell's paradox by making redundant the question of whether or not to enter the catalogue of catalogues which do not list themselves in itself.

Russell spent the next decade considering the axioms of mathematics, the very essence of the subject. Then in 1910, in partnership with Alfred North Whitehead, he published the first of three volumes of *Principia Mathematica* – an apparently successful attempt to partly address the problem created by his own paradox. For the next two decades others used *Principia Mathematica* as a guide for establishing a flawless mathematical edifice, and by the time Hilbert retired in 1930 he felt confident that mathematics was well on the road to recovery. His dream of a consistent logic, powerful enough to answer every question, was apparently on its way to becoming a reality.

Then in 1931 an unknown twenty-five-year-old mathematician published a paper which would forever destroy Hilbert's hopes. Kurt Gödel would force mathematicians to accept that mathematics could never be logically perfect, and implicit in his works was the idea that problems like Fermat's Last Theorem might even be impossible to solve.

Kurt Gödel was born on 28 April 1906 in Moravia, then part of the Austro-Hungarian Empire, now part of the Czech Republic. From an early age he suffered from severe illness, the most serious being a bout of rheumatic fever at the age of six. This early brush with death caused Gödel to develop an obsessive hypochondria which stayed with him throughout his life. At the age of eight, having read a medical textbook, he became convinced that he had a weak heart, even though his doctors could find no evidence of the condition. Later, towards the end of his life, he mistakenly believed that he was being poisoned and refused to eat, almost starving himself to death.

As a child Gödel displayed a talent for science and mathematics, and his inquisitive nature led his family to nickname him *der Herr Warum* (Mr Why). He went to the University of Vienna unsure of whether to specialise in mathematics or physics, but an inspiring and passionate lecture course on number theory by Professor P. Furtwängler persuaded Gödel to devote his life to numbers. The lectures were all the more extraordinary because Furtwängler was paralysed from the neck down and had to lecture from his wheelchair without notes while his assistant wrote on the blackboard.

By his early twenties Gödel had established himself in the mathematics department, but along with his colleagues he would occasionally wander down the corridor to attend meetings of the *Wiener Kreis* (Viennese Circle), a group of philosophers who would gather to discuss the day's great questions of logic. It was during this period that Gödel developed the ideas that would devastate the foundations of mathematics.

In 1931 Gödel published his book *Über formal unentscheidbare Sätze der Principia Mathematica und verwandter Systeme* (On Formally Undecidable Propositions in *Principia Mathematica* and Related Systems), which contained his so-called theorems of undecidability. When news of the theorems reached America the great mathematician John von Neumann immediately cancelled a lecture series he was giving on Hilbert's programme and replaced the

Kurt Gödel

remainder of the course with a discussion of Gödel's revolutionary work.

Gödel had proved that trying to create a complete and consistent mathematical system was an impossible task. His ideas could be encapsulated in two statements.

First theorem of undecidability
If axiomatic set theory is consistent, there exist theorems which can neither be proved or disproved.

Second theorem of undecidability
There is no constructive procedure which will prove axiomatic theory to be consistent.

Essentially Gödel's first statement said that no matter what set of axioms were being used there would be questions which mathematics could not answer – completeness could never be achieved. Worse still, the second statement said that mathematicians could never even be sure that their choice of axioms would not lead to a contradiction – consistency could never be proved. Gödel had shown that the Hilbert programme was an impossible exercise.

Decades later, in *Portraits from Memory*, Bertrand Russell reflected on his reaction to Gödel's discovery:

I wanted certainty in the kind of way in which people want religious faith. I thought that certainty is more likely to be found in mathematics than elsewhere. But I discovered that many mathematical demonstrations, which my teachers expected me to accept, were full of fallacies, and that, if certainty were indeed discoverable in mathematics, it would be in a new field of mathematics, with more solid foundations than those that had hitherto been thought secure. But as the work proceeded, I was continually reminded of the fable about the elephant and the tortoise. Having constructed an elephant upon which the mathematical world could rest,

I found the elephant tottering, and proceeded to construct a tortoise to keep the elephant from falling. But the tortoise was no more secure than the elephant, and after some twenty years of arduous toil, I came to the conclusion that there was nothing more that I could do in the way of making mathematical knowledge indubitable.

Although Gödel's second statement said that it was impossible to prove that the axioms were consistent, this did not necessarily mean that they were inconsistent. In their hearts many mathematicians still believed that their mathematics would remain consistent, but in their minds they could not prove it. Many years later the great number theorist André Weil would say: 'God exists since mathematics is consistent, and the Devil exists since we cannot prove it.'

The proof of Gödel's theorems of undecidability is immensely complicated, and in fact a more rigorous statement of the First Theorem should be:

To every ω-consistent recursive class κ of *formulae* there correspond recursive *class-signs* r, such that neither v Gen r nor Neg(v Gen r) belongs to Flg(κ) (where v is the *free variable* of r).

Fortunately, as with Russell's paradox and the tale of the librarian, Gödel's first theorem can be illustrated with another logical analogy due to Epimenides and known as the *Cretan paradox*, or *liar's paradox*. Epimenides was a Cretan who exclaimed:

'*I am a liar!*'

The paradox arises when we try and determine whether this statement is true or false. First let us see what happens if we assume that the statement is true. A true statement implies that Epimenides is a liar, but we initially assumed that he made a true statement and therefore Epimenides is not a liar – we have an inconsistency. On the other hand let us see what happens if we assume that the statement is false. A false statement implies that Epimenides is not a liar, but we initially assumed that he made a false statement and therefore Epimenides is a liar – we have another inconsistency. Whether we assume that the statement is true or false we end up with an inconsistency, and therefore the statement is neither true nor false.

Gödel reinterpreted the liar's paradox and introduced the concept of proof. The result was a statement along the following lines:

This statement does not have any proof.

If the statement were false then the statement would be provable, but this would contradict the statement. Therefore the statement must be true in order to avoid the contradiction. However, although the statement is true it cannot be proven, because this statement (which we now know to be true) says so.

Because Gödel could translate the above statement into mathematical notation, he was able to demonstrate that there existed statements in mathematics which are true but which could never be proven to be true, so-called undecidable statements. This was the death-blow for the Hilbert programme.

In many ways Gödel's work paralleled similar discoveries being made in quantum physics. Just four years before Gödel published his work on undecidability, the German physicist Werner Heisenberg uncovered the uncertainty principle. Just as there was a fundamental limit to what theorems mathematicians could prove, Heisenberg showed that there was a fundamental limit to what properties physicists could measure. For example, if they wanted to measure the exact position of an object, then they could measure the object's velocity with only relatively poor accuracy. This is because in order to measure the position of the object it would be necessary to illuminate it with photons of light, but to pinpoint its exact locality the photons of light would have to have enormous energy. However, if the object is being bombarded by high-energy photons its own velocity will be affected and becomes inherently uncertain. Hence, by demanding knowledge of an object's position, physicists would have to give up some knowledge of its velocity.

Heisenberg's uncertainty principle only reveals itself at atomic scales, when high-precision measurements become critical. Therefore much of physics could carry on regardless while quantum physicists concerned themselves with profound questions about the limits of knowledge. The same was happening in the world of mathematics. While the logicians concerned themselves with a highly esoteric debate about undecidability, the rest of the mathematical community carried on regardless. Although Gödel had proved that there were some statements which could not be proven, there were plenty of statements which could be proven and his discovery did not invalidate anything proven in the past. Furthermore, many mathematicians believed that Gödel's undecidable statements would only be found in the most obscure and extreme regions of mathematics and might therefore never be encountered. After all Gödel had only said that these statements existed; he could not actually point to one. Then in 1963 Gödel's theoretical nightmare became a full-blooded reality.

Paul Cohen, a twenty-nine-year-old mathematician at Stanford University, developed a technique for testing whether or not a particular question is undecidable. The technique only works in a few very special cases, but he was nevertheless the first person to discover specific questions which were indeed undecidable. Having made his discovery Cohen immediately flew to Princeton, proof in hand, to have it verified by Gödel himself. Gödel, who by now was entering a paranoid phase of his life, opened the door slightly, snatched the papers and slammed the door shut. Two days later Cohen received an invitation to tea at Gödel's house, a sign that the master had given the proof his stamp of authority. What was particularly dramatic was that some of these undecidable questions were central to mathematics. Ironically Cohen proved that one of the questions which David Hilbert declared to be among the twenty-three most important problems in mathematics, the continuum hypothesis, was undecidable.

Gödel's work, compounded by Cohen's undecidable statements, sent a disturbing message to all those mathematicians, professional and amateur, who were persisting in their attempts to prove Fermat's Last Theorem – perhaps Fermat's Last Theorem was undecidable! What if Pierre de Fermat had made a mistake when he claimed to have found a proof? If so, then there was the possibility that the Last Theorem was undecidable. Proving Fermat's Last Theorem might be more than just difficult, it might be impossible. If Fermat's Last Theorem were undecidable, then mathematicians had spent centuries in search of a proof that did not exist.

Curiously if Fermat's Last Theorem turned out to be undecidable, then this would imply that it must be true. The reason is as follows. The Last Theorem says that there are no whole number solutions to the equation

$$x^n + y^n = z^n \quad \text{for } n \text{ greater than } 2.$$

If the Last Theorem were in fact false, then it would be possible to prove this by identifying a solution (a counter-example). Therefore the Last Theorem would be decidable. Being false would be inconsistent with being undecidable. However, if the Last Theorem were true, there would not necessarily be such an unequivocal way of proving it so, i.e. it could be undecidable. In conclusion, Fermat's Last Theorem might be true, but there may be no way of proving it.

The Compulsion of Curiosity

Pierre de Fermat's casual jotting in the margin of Diophantus' *Arithmetica* had led to the most infuriating riddle in history. Despite three centuries of glorious failure and Gödel's suggestion that they might be hunting for a non-existent proof, some mathematicians continued to be attracted to the problem. The Last Theorem was a mathematical siren, luring geniuses towards it, only to dash their hopes. Any mathematician who got involved with Fermat's Last Theorem risked wasting their career, and yet whoever could make the crucial breakthrough would go down in history as having solved the world's most difficult problem.

Generations of mathematicians were obsessed with Fermat's Last Theorem for two reasons. First, there was the ruthless sense of one-upmanship. The Last Theorem was the ultimate test and who-ever could prove it would succeed where Cauchy, Euler, Kummer, and countless others had failed. Just as Fermat himself took great pleasure in solving problems which baffled his contemporaries, whoever could prove the Last Theorem could enjoy the fact that they had solved a problem which had confounded the entire com-munity of mathematicians for hundreds of years. Second, whoever could meet Fermat's challenge could enjoy the innocent satisfac-tion of solving a riddle. The delight derived from solving esoteric questions in number theory is not so different from the simple joy of tackling the trivial riddles of Sam Loyd. A mathematician once said to me that the pleasure he derived from solving mathematical problems is similar to that gained by crossword addicts. Filling in the last clue of a particularly tough crossword is always a satisfying experience, but imagine the sense of achievement after spending years on a puzzle, which nobody else in the world has been able to solve, and then figuring out the solution.

These are the same reasons why Andrew Wiles became fascinated by Fermat: 'Pure mathematicians just love a challenge. They love unsolved problems. When doing maths there's this great feeling. You start with a problem that just mystifies you. You can't understand it, it's so complicated, you just can't make head nor tail of it. But then when you finally resolve it, you have this incredible feeling of how beautiful it is, how it all fits together so elegantly. Most deceptive are the problems which look easy, and yet they turn out to be extremely intricate. Fermat is the most beautiful example of this. It just looked as though it had to have a solution and, of course, it's very special because Fermat said that he had a solution.'

Mathematics has its applications in science and technology, but that is not what drives mathematicians. They are inspired by the joy of discovery. G.H. Hardy tried to explain and justify his own career in a book entitled *A Mathematician's Apology*:

I will only say that if a chess problem is, in the crude sense, 'useless', then that is equally true of most of the best mathematics . . . I have never done anything 'useful'. No discovery of mine has made, or is likely to make, directly or indirectly, for good or ill, the least difference to the amenity of the world. Judged by all practical standards, the value of my mathematical life is nil; and outside mathematics it is trivial anyhow. I have just one chance of escaping a verdict of complete triviality, that I may be judged to have created something worth creating. And that I have created something is undeniable: the question is about its value.

The desire for a solution to any mathematical problem is largely fired by curiosity, and the reward is the simple but enormous satisfaction derived from solving any riddle. The mathematician E.C. Titchmarsh once said: 'It can be of no practical use to know that π is irrational, but if we can know, it surely would be intolerable not to know.'

In the case of Fermat's Last Theorem there was no shortage of curiosity. Gödel's work on undecidability had introduced an element of doubt as to whether the problem was soluble, but this was not enough to discourage the true Fermat fanatic. What was more dispiriting was the fact that by the 1930s mathematicians had exhausted all their techniques and had little else at their disposal. What was needed was a new tool, something that would raise mathematical morale. The Second World War was to provide just what was required – the greatest leap in calculating power since the invention of the slide-rule.

The Brute Force Approach

When in 1940 G.H. Hardy declared that the best mathematics is largely useless, he was quick to add that this was not necessarily a bad thing: 'Real mathematics has no effects on war. No one has yet discovered any warlike purpose to be served by the theory of numbers.' Hardy was soon to be proved wrong.

In 1944 John von Neumann co-wrote the book *The Theory of Games and Economic Behavior*, in which he coined the term *game theory*. Game theory was von Neumann's attempt to use mathematics to describe the structure of games and how humans play them. He began by studying chess and poker, and then went on to try and model more sophisticated games such as economics. After the

Second World War the RAND corporation realised the potential of von Neumann's ideas and hired him to work on developing Cold War strategies. From that point on, mathematical game theory has become a basic tool for generals to test their military strategies by treating battles as complex games of chess. A simple illustration of the application of game theory in battles is the story of the *truel*.

A truel is similar to a duel, except there are three participants rather than two. One morning Mr Black, Mr Grey and Mr White decide to resolve a conflict by truelling with pistols until only one of them survives. Mr Black is the worst shot, hitting his target on average only one time in three. Mr Grey is a better shot hitting his target two times out of three. Mr White is the best shot hitting his target every time. To make the truel fairer Mr Black is allowed to shoot first, followed by Mr Grey (if he is still alive), followed by Mr White (if he is still alive), and round again until only one of them is alive. The question is this: Where should Mr Black aim his first shot? You might like to make a guess based on intuition, or better still based on game theory. The answer is discussed in Appendix 9.

Even more influential in wartime than game theory is the mathematics of code breaking. During the Second World War the Allies realised that in theory mathematical logic could be used to unscramble German messages, if only the calculations could be performed quickly enough. The challenge was to find a way of automating mathematics so that a machine could perform the calculations, and the Englishman who contributed most to this code-cracking effort was Alan Turing.

In 1938 Turing returned to Cambridge having completed a stint at Princeton University. He had witnessed first-hand the turmoil caused by Gödel's theorems of undecidability and had become involved in trying to pick up the pieces of Hilbert's dream. In particular he wanted to know if there was a way to define which questions were and were not decidable, and tried to develop a methodical way of answering this question. At the time calculating devices were primitive and effectively useless when it came to serious mathematics, and so instead Turing based his ideas on the concept of an imaginary machine which was capable of infinite computation. This hypothetical machine, which consumed infinite amounts of imaginary ticker-tape and could compute for an eternity, was all that he required to explore his abstract questions of logic. What Turing was unaware of was that his imagined mechanisation of hypothetical questions would eventually lead to a breakthrough in performing real calculations on real machines.

Despite the outbreak of war, Turing continued his research as a fellow of King's College, until on 4 September 1940 his contented life as a Cambridge don came to an abrupt end. He had been commandeered by the Government Code and Cypher School, whose task it was to unscramble the enemy's coded messages. Prior to the war the Germans had devoted considerable effort to developing a superior system of encryption, and this was a matter of grave concern to British Intelligence who had in the past been able to decipher their enemy's communications with relative ease. The HMSO's official war history *British Intelligence in the Second World War* describes the state of play in the 1930s:

By 1937 it was established that, unlike their Japanese and Italian counterparts, the German army, the German navy and probably the air force, together with other state organisations like the railways and the SS used, for all except their tactical communications, different versions of the same cypher system – the Enigma machine which had been put on the market in the 1920s but which the Germans had rendered more secure by progressive modifications. In 1937 the Government Code and Cypher School broke into the less modified and less secure model of this machine

Alan Turing

that was being used by the Germans, the Italians and the Spanish nation-
alist forces. But apart from this the Enigma still resisted attack, and it
seemed likely that it would continue to do so.

The Enigma machine consisted of a keyboard connected to a
scrambler unit. The scrambler unit contained three separate rotors
and the positions of the rotors determined how each letter on the
keyboard would be enciphered. What made the Enigma code so
difficult to crack was the enormous number of ways in which the
machine could be set up. First, the three rotors in the machine
were chosen from a selection of five, and could be changed and
swapped around to confuse the code-breakers. Second, each rotor
could be positioned in one of twenty-six different ways. This means
that the machine could be set up in over a million different ways.
In addition to the permutations provided by the rotors, plugboard
connections at the back of the machine could be changed by hand
to provide a total of over 150 million million million possible set-
ups. To increase security even further, the three rotors were continually changing their orientation, so that every time a letter was
transmitted, the set-up for the machine, and therefore the enci-
pherment, would change for the next letter. So typing 'DODO'
could generate the message 'FGTB' – the 'D' and the 'O' are sent
twice, but encoded differently each time.

Enigma machines were given to the German army, navy and air
force, and were even operated by the railways and other government departments. As with all code systems used during this
period, a weakness of the Enigma was that the receiver had to
know the sender's Enigma setting. To maintain security the
Enigma settings had to be changed on a daily basis. One way for
senders to change settings regularly and keep receivers informed
was to publish the daily settings in a secret code-book. The risk

with this approach was that the British might capture a U-boat and obtain the code-book with all the daily settings for the following month. The alternative approach, and the one adopted for the bulk of the war, was to transmit the daily settings in a preamble to the actual message, encoded using the previous day's settings.

When the war started, the British Cypher School was dominated by classicists and linguists. The Foreign Office soon realised that number theorists had a better chance of finding the key to cracking the German codes and, to begin with, nine of Britain's most brilliant number theorists were gathered at the Cypher School's new home at Bletchley Park, a Victorian mansion in Bletchley, Buckinghamshire. Turing had to abandon his hypothetical machines with infinite ticker-tape and endless processing time and come to terms with a practical problem with finite resources and a very real deadline.

Cryptography is an intellectual battle between the code-maker and the code-breaker. The challenge for the code-maker is to shuffle and scramble an outgoing message to the point where it would be indecipherable if intercepted by the enemy. However, there is a limit on the amount of mathematical manipulation possible because of the need to dispatch messages quickly and efficiently. The strength of the German Enigma code was that the coded message underwent several levels of encryption at very high speed. The challenge for the code-breaker was to take an intercepted message and to crack the code while the contents of the message were still relevant. A German message ordering a British ship to be destroyed had to be decoded before the ship was sunk.

Turing led a team of mathematicians who attempted to build mirror-images of the Enigma machine. Turing incorporated his pre-war abstract ideas into these devices, which could in theory methodically check all the possible Enigma machine set-ups until the code was cracked. The British machines, over two metres tall and equally wide, employed electromechanical relays to check all the potential Enigma settings. The constant ticking of the relays led to them being nicknamed *bombes*. Despite their speed it was impossible for the bombes to check every one of the 150 million million million possible Enigma settings within a reasonable amount of time, and so Turing's team had to find ways to significantly reduce the number of permutations by gleaning whatever information they could from the sent messages.

One of the greatest breakthroughs made by the British was the realisation that the Enigma machine could never encode a letter into itself, i.e. if the sender tapped 'R' then the machine could potentially send out any letter, depending on the settings of the machine, apart from 'R'. This apparently innocuous fact was all that was needed to drastically reduce the time required to decipher a message. The Germans fought back by limiting the length of the messages they sent. All messages inevitably contain clues for the team of code-breakers, and the longer the message, the more clues it contains. By limiting all messages to a maximum of 250 letters, the Germans hoped to compensate for the Enigma machine's reluctance to encode a letter as itself.

In order to crack codes Turing would often try to guess key-words in messages. If he was right it would speed up enormously the cracking of the rest of the code. For example, if the code-breakers suspected that a message contained a weather report, a frequent type of coded report, then they would guess that the message contained words such as 'fog' or 'windspeed'. If they were right they could quickly crack that message, and thereby deduce the Enigma settings for that day. For the rest of the day, other, more valuable, messages could be broken with ease.

When they failed to guess weather words, the British would try and put themselves in the position of the German Enigma operators to guess other keywords. A sloppy operator might address the receiver by a first name or he might have developed idiosyncrasies which were known to the code-breakers. When all else failed and German traffic was flowing unchecked, it is said that the British Cypher School even resorted to asking the RAF to mine a particular German harbour. Immediately the German harbour-master would send an encrypted message which would be intercepted by the British. The code-breakers could be confident that the message contained words like 'mine', 'avoid' and 'map reference'. Having cracked this message, Turing would have that day's Enigma settings and any further German traffic was vulnerable to rapid decipherment.

On 1 February 1942 the Germans added a fourth wheel to Enigma machines which were employed for sending particularly sensitive information. This was the greatest escalation in the level of encryption during the war, but eventually Turing's team fought back by increasing the efficiency of the bombes. Thanks to the Cypher School, the Allies knew more about their enemy than the Germans could ever have suspected. The impact of German U-boats in the Atlantic was greatly reduced and the British had advanced warning of attacks by the Luftwaffe. The code-breakers also intercepted and deciphered the exact position of German supply ships, allowing British destroyers to be sent out to sink them.

At all times the Allied forces had to take care that their evasive actions and uncanny attacks did not betray their ability to decipher German communications. If the Germans suspected that Enigma had been broken then they would increase their level of encryption, and the British might be back to square one. Hence there were occasions when the Cypher School informed the Allies of an imminent attack, and the Allies chose not to take extreme countermeasures. There are even rumours that Churchill knew that Coventry was to be targeted for a devastating raid, yet he chose not to take special precautions in case the Germans became suspicious. Stuart Milner-Barry who worked with Turing denies the rumour, and states that the relevant message concerning Coventry was not cracked until it was too late.

The restrained use of decoded information worked perfectly. Even when the British used intercepted communications to inflict heavy losses, the Germans did not suspect that the Enigma code had been broken. They believed that their level of encryption was so high that it would be absolutely impossible to crack their codes. Instead they blamed any exceptional losses on the British secret service infiltrating their own ranks.

Because of the secrecy surrounding the work carried out at Bletchley by Turing and his team, their immense contribution to the war effort could never be publicly acknowledged, even for many years after the war. It used to be said that the First World War was the chemists' war and that the Second World War was the physicists' war. In fact, from the information revealed in recent decades, it is probably true to say that the Second World War was also the mathematicians' war – and in the case of a third world war their contribution would be even more critical.

Throughout his code-breaking career Turing never lost sight of his mathematical goals. The hypothetical machines had been replaced with real ones, but the esoteric questions remained. By the end of the war Turing had helped build Colossus, a fully electronic machine consisting of 1500 valves, which were much faster than the electromechanical relays employed in the bombes. Colossus was a computer in the modern sense of the word, and with the extra speed and sophistication Turing began to think of it as a primitive brain – it had a memory, it could process information, and states within the computer resembled states of mind. Turing had transformed his imaginary machine into the first real computer.

When the war ended Turing continued to build increasingly complex machines, such as the Automatic Computing Engine (ACE). In 1948 he moved to Manchester University and built the world's first computer to have an electronically stored program. Turing had provided Britain with the most advanced computers in the world, but he would not live long enough to see their most remarkable calculations.

In the years after the war Turing had been under surveillance from British Intelligence, who were aware that he was a practising homosexual. They were concerned that the man who knew more about Britain's security codes than anyone else was vulnerable to blackmail and decided to monitor his every move. Turing had largely come to terms with being constantly shadowed, but in 1952 he was arrested for violation of British homosexuality statutes. This humiliation made life intolerable for Turing. Andrew Hodges, Turing's biographer, describes the events leading up to his death:

Alan Turing's death came as a shock to those who knew him . . . That he was an unhappy, tense, person; that he was consulting a psychiatrist and suffered a blow that would have felled many people – all this was clear. But the trial was two years in the past, the hormone treatment had ended a year before, and he seemed to have risen above it all.

The inquest, on 10 June 1954, established that it was suicide. He had been found lying neatly in his bed. There was froth round his mouth, and the pathologist who did the post-mortem easily identified the cause of death as cyanide poisoning . . . In the house was a jar of potassium cyanide, and also a jar of cyanide solution. By the side of his bed was half an apple, out of which several bites had been taken. They did not analyse the apple, and so it was never properly established that, as seemed perfectly obvious, the apple had been dipped in the cyanide.

Turing's legacy was a machine which could take an impractically long calculation, if performed by a human, and complete it in a matter of hours. Today's computers perform more calculations in a split second than Fermat performed in his entire career. Mathematicians who were still struggling with Fermat's Last Theorem began to use computers to attack the problem, relying on a computerised version of Kummer's nineteenth-century approach.

Kummer, having discovered a flaw in the work of Cauchy and Lamé, showed that the outstanding problem in proving Fermat's Last Theorem was disposing of the cases when n equals an irregular prime – for values of n up to 100 the only irregular primes are 37, 59 and 67. At the same time Kummer showed that in theory all irregular primes could be dealt with on an individual basis, the only problem being that each one would require an enormous amount of calculation. To make his point Kummer and his colleague Dimitri Mirimanoff put in the weeks of calculation required to dispel the three irregular primes less than 100. However, they and other mathematicians were not prepared to begin on the next batch of irregular primes between 100 and 1,000.

A few decades later the problems of immense calculation began to vanish. With the arrival of the computer awkward cases of Fermat's Last Theorem could be dispatched with speed, and after the Second World War teams of computer scientists and mathematicians proved Fermat's Last Theorem for all values of n up to 500, then 1,000, and then 10,000. In the 1980s Samuel S. Wagstaff of the University of Illinois raised the limit to 25,000 and more recently mathematicians could claim that Fermat's Last Theorem was true for all values of n up to 4 million.

Although outsiders felt that modern technology was at last getting the better of the Last Theorem, the mathematical community were aware that their success was purely cosmetic. Even if supercomputers spent decades proving one value of n after another they could never prove every value of n up to infinity, and therefore they could never claim to prove the entire theorem. Even if the theorem was to be proved true for up to a billion, there is no reason why it should be true for a billion and one. If the theorem was to be proved up to a trillion, there is no reason why it should be true for a trillion and one, and so on *ad infinitum*. Infinity is unobtainable by the mere brute force of computerised number crunching.

David Lodge in his book *The Picturegoers* gives a beautiful description of eternity which is also relevant to the parallel concept of infinity: 'Think of a ball of steel as large as the world, and a fly alighting on it once every million years. When the ball of steel is rubbed away by the friction, eternity will not even have begun.'

All that computers could offer was evidence in favour of Fermat's Last Theorem. To the casual observer the evidence might seem to be overwhelming, but no amount of evidence is enough to satisfy mathematicians, a community of sceptics who will accept nothing other than absolute proof. Extrapolating a theory to cover an infinity of numbers based on evidence from a few numbers is a risky (and unacceptable) gamble.

One particular sequence of primes shows that extrapolation is a dangerous crutch upon which to rely. In the seventeenth century mathematicians showed by detailed examination that the following numbers are all prime:

31; 331; 3,331; 33,331; 333,331; 3,333,331; 33,333,331.

The next numbers in the sequence become increasingly giant, and checking whether or not they are also prime would have taken considerable effort. At the time some mathematicians were tempted to extrapolate from the pattern so far, and assume that all numbers of this form are prime. However, the next number in the pattern, 333,333,331, turned out not to be a prime:

$$333,333,331 = 17 \times 19,607,843.$$

Another good example which demonstrates why mathematicians refused to be persuaded by the evidence of computers is the case of Euler's conjecture. Euler claimed that there were no solutions to an equation not dissimilar to Fermat's equation:

$$x^4 + y^4 + z^4 = w^4.$$

For two hundred years nobody could prove Euler's conjecture, but on the other hand nobody could disprove it by finding a counter-example. First manual searches and then years of computer sifting failed to find a solution. Lack of a counter-example was strong evidence in favour of the conjecture. Then in 1988 Naom Elkies of Harvard University discovered the following solution:

$$2,682,440^4 + 15,365,639^4 + 18,796,760^4 = 20,615,673^4.$$

Despite all the evidence Euler's conjecture turned out to be false. In fact Elkies proved that there were infinitely many solutions to the equation. The moral is that you cannot use evidence from the first million numbers to prove a conjecture about all numbers.

But the deceptive nature of Euler's conjecture is nothing compared to the *overestimated prime conjecture*. By scouring through larger and larger regimes of numbers, it becomes clear that the prime numbers become harder and harder to find. For instance, between 0 and 100 there are 25 primes but between 10,000,000 and 10,000,100 there are only 2 prime numbers. In 1791, when he was just fourteen years old, Carl Gauss predicted the approximate manner in which the frequency of prime numbers among all the other numbers would diminish. The formula was reasonably accurate but always seemed slightly to overestimate the true distribution of primes. Testing for primes up to a million, a billion or a trillion would always show that Gauss's formula was marginally too generous and mathematicians were strongly tempted to believe that this would hold true for all numbers up to infinity, and thus was born the overestimated prime conjecture.

Then, in 1914, J.E. Littlewood, G.H. Hardy's collaborator at Cambridge, proved that in a sufficiently large regime Gauss's formula would *underestimate* the number of primes. In 1955 S. Skewes showed that the underestimate would occur sometime before reaching the number

$$10^{10^{10,000,000,000,000,000,000,000,000,000,000,000}}$$

This is a number beyond the imagination, and beyond any practical application. Hardy called Skewes's number 'the largest number which has ever served any definite purpose in mathematics'. He calculated that if one played chess with all the particles in the universe (10^{87}), where a move meant simply interchanging any two particles, then the number of possible games was roughly Skewes's number.

There was no reason why Fermat's Last Theorem should not turn out to be as cruel and deceptive as Euler's conjecture or the overestimated prime conjecture.

The Graduate

In 1975 Andrew Wiles began his career as a graduate student at Cambridge University. Over the next three years he was to work on his Ph.D. thesis and in that way serve his mathematical apprenticeship. Each student was guided and nurtured by a supervisor and in Wiles's case that was the Australian John Coates, a professor at Emmanuel College, originally from Possum Brush, New South Wales.

Coates still recalls how he adopted Wiles: 'I remember a colleague told me that he had a very good student who was just finishing part III of the mathematical tripos, and he urged me to take him as a student. I was very fortunate to have Andrew as a student. Even as a research student he had very deep ideas and it was always clear that he was a mathematician who would do great things. Of course, at that stage there was no question of any research student starting work directly on Fermat's Last Theorem. It was too difficult even for a thoroughly experienced mathematician.'

For the past decade everything Wiles had done was directed towards preparing himself to meet Fermat's challenge, but now that he had joined the ranks of the professional mathematicians he had to be more pragmatic. He remembers how he had to temporarily surrender his dream: 'When I went to Cambridge I really put aside Fermat. It's not that I forgot about it – it was always there – but I realised that the only techniques we had to tackle it had been around for 130 years. It didn't seem that these techniques were really getting to the root of the problem. The problem with working on Fermat was that you could spend years getting nowhere. It's fine to work on any problem, so long as it generates interesting mathematics along the way – even if you don't solve it at the end of the day. The definition of a good mathematical problem is the mathematics it generates rather than the problem itself.'

It was John Coates's responsibility to find Andrew a new obsession, something which would occupy his research for at least the

Andrew Wiles during his college years.

John Coates, Wiles's supervisor in the 1970s, has continued to keep in contact with his former student.

next three years. 'I think all a research supervisor can do for a student is try and push him in a fruitful direction. Of course, it's impossible to be sure what is a fruitful direction in terms of research but perhaps one thing that an older mathematician can do is use his horse sense, his intuition of what is a good area, and then it's really up to the student as to how far he can go in that direction.' In the end Coates decided that Wiles should study an area of mathematics known as *elliptic curves*. This decision would eventually prove to be a turning point in Wiles's career and give him the techniques he would require for a new approach to tackling Fermat's Last Theorem.

The name 'elliptic curves' is somewhat misleading for they are neither ellipses nor even curved in the normal sense of the word. Rather they are any equations which have the form

$y^2 = x^3 + ax^2 + bx + c,$ where a, b, c are any whole numbers.

They got their name because in the past they were used to measure the perimeters of ellipses and the lengths of planetary orbits, but for clarity I will simply refer to them as *elliptic equations* rather than elliptic curves.

The challenge with elliptic equations, as with Fermat's Last Theorem, is to figure out if they have whole number solutions, and, if so, how many. For example, the elliptic equation

$$y^2 = x^3 - 2,   \text{where } a = 0, b = 0, c = -2,$$

has only one set of whole number solutions, namely

$$5^2 = 3^3 - 2,   \text{or}   25 = 27 - 2.$$

Proving that this elliptic equation has only one set of whole number solutions is an immensely difficult task, and in fact it was Pierre de Fermat who discovered the proof. You might remember that in Chapter 2 it was Fermat who proved that 26 is the only number in the universe sandwiched between a square and a cube number. This is equivalent to showing that the above elliptic equation has only one solution, i.e. 5^2 and 3^3 are the only square and cube that differ by 2, and therefore 26 is the only number that can be sandwiched between two such numbers.

What makes elliptic equations particularly fascinating is that they occupy a curious niche between other simpler equations which are almost trivial and other more complicated equations which are impossible to solve. By simply changing the values of a, b and c in the general elliptic equation mathematicians can generate an infinite variety of equations, each one with its own characteristics, but all of them just within the realm of solubility.

Elliptic equations were originally studied by the ancient Greek mathematicians, including Diophantus who devoted large parts of his *Arithmetica* to exploring their properties. Probably inspired by Diophantus, Fermat also took up the challenge of elliptic equations, and, because they had been studied by his hero, Wiles was happy to explore them further. Even after two thousand years elliptic equations still offered formidable problems for students such as Wiles: 'They are very far from being completely understood. There are many apparently simple questions I could pose on elliptic equations that are still unresolved. Even questions that Fermat himself considered are still unresolved. In some way all the mathematics that I've done can trace its ancestry to Fermat, if not Fermat's Last Theorem.'

In the equations which Wiles studied as a graduate student, determining the exact number of solutions was so difficult that the only way to make any progress was to simplify the problem. For example, the following elliptic equation is almost impossible to tackle directly:

$$x^3 - x^2 = y^2 + y.$$

The challenge is to figure out how many whole number solutions there are to the equation. One fairly trivial solution is $x = 0$ and $y = 0$:

$$0^3 - 0^2 = 0^2 + 0.$$

A slightly more interesting solution is $x = 1$ and $y = 0$:

$$1^3 - 1^2 = 0^2 + 0.$$

There may be other solutions but, with an infinite quantity of whole numbers to investigate, giving a complete list of solutions to this particular equation is an impossible task. A simpler task is to look for solutions within a finite number space, so-called clock arithmetic.

Earlier we saw how numbers can be thought of as marks along the number line which extends to infinity, as shown in Figure 16. To make the number space finite, clock arithmetic involves truncating the line and looping it back on itself to form a number ring as opposed to a number line. Figure 17 shows a 5-clock, where the number line has been truncated at 5 and looped back to 0. The number 5 vanishes and becomes equivalent to 0, and therefore the only numbers in 5-clock arithmetic are 0, 1, 2, 3, 4.

In normal arithmetic we can think of addition as moving along the line a certain number of spaces. For example, $4 + 2 = 6$ is the same as saying: begin at 4, and move along the number line 2 spaces, and arrive at 6.

However, in 5-clock arithmetic:

$$4 + 2 = 1.$$

This is because if we start at 4 and move round 2 spaces then we arrive back at 1. Clock arithmetic might appear unfamiliar but in

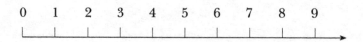

Figure 16. Conventional arithmetic can be thought of as movements up and down the number line.

fact, as the name suggests, it is used every day when people discuss the time. Four hours after 11 o'clock (that is to say, 11 + 4) is generally not called 15 o'clock, but rather 3 o'clock. This is 12-clock arithmetic.

As well as addition we can perform all the other common mathematical operations, such as multiplication. In 12-clock arithmetic 5 × 7 = 11. This multiplication can be thought of as follows: if you start at 0, then move along 5 lots of 7 spaces, you will eventually arrive at 11. Although this is one way of thinking about multiplication in clock arithmetic, there are short cuts which speed up calculations. For example, to calculate 5 × 7 in 12-clock arithmetic, we can begin by just working out the normal result which is 35. We then divide 35 by 12 and work out the remainder, which is the answer to the original question. So 12 goes into 35 only twice, with a remainder of 11, and sure enough 5 × 7 in 12-clock arithmetic is 11. This is equivalent to imagining going around the clock twice and still having 11 spaces to travel.

Because clock arithmetics only deal with a limited number space, it is relatively easy to work out all the possible solutions to an elliptic equation for a given clock arithmetic. For example, working in 5-clock arithmetic it is possible to list all the possible solutions to the elliptic equation

$$x^3 - x^2 = y^2 + y.$$

Figure 17. In 5-clock arithmetic the number line is truncated at 5 and looped back on itself. The number 5 coincides with 0, and therefore is replaced by it.

The solutions are:

$$x = 0, \qquad y = 0,$$
$$x = 0, \qquad y = 4,$$
$$x = 1, \qquad y = 0,$$
$$x = 1, \qquad y = 4.$$

Although some of these solutions would not be valid in normal arithmetic, in 5-clock arithmetic they are acceptable. For example, the fourth solution ($x = 1$, $y = 4$) works as follows:

$$x^3 - x^2 = y^2 + y$$
$$1^3 - 1^2 = 4^2 + 4$$
$$1 - 1 = 16 + 4$$
$$0 = 20.$$

But remember, 20 is equivalent to 0 in 5-clock arithmetic, because 5 will divide into 20 with a remainder of 0.

Because they could not list all the solutions to an elliptic equation working in infinite space, mathematicians, including Wiles, settled for working out the number of solutions in all the different clock arithmetics. For the elliptic equation given above the number of solutions in 5-clock arithmetic is four, and so mathematicians say $E_5 = 4$. The number of solutions in other clock arithmetics can also be calculated. For example, in 7-clock arithmetic the number of solutions is nine, and so $E_7 = 9$.

To summarise their results, mathematicians list the number of solutions in each clock arithmetic and call this list the *L*-series for the elliptic equation. What the *L* stands for has been long forgotten although some have suggested that it is the *L* of Gustav Lejeune-Dirichlet, who worked on elliptic equations. For clarity I will use the term *E*-series – the series that is derived from an elliptic equation. For the example given above the *E*-series is as follows:

Elliptic equation: $x^3 - x^2 = y^2 + y$;

E-series: $E_1 = 1,$

$E_2 = 4,$

$E_3 = 4,$

$E_4 = 8,$

$E_5 = 4,$

$E_6 = 16,$

$E_7 = 9,$

$E_8 = 16,$

$\vdots$

Because mathematicians cannot say how many solutions some elliptic equations have in normal number space which extends up to infinity, the *E*-series appears to be next best thing. In fact the *E*-series encapsulates a great deal of information about the elliptic equation it describes. In the same way that biological DNA carries all the information required to construct a living organism, the E-series carries the essence of the elliptic equation. The hope was that by studying the E-series, this mathematical DNA, mathematicians would ultimately be able to calculate everything they could ever wish to know about an elliptic equation.

Working alongside John Coates, Wiles rapidly established his reputation as a brilliant number theorist with a profound under-standing of elliptic equations and their E-series. As each new result was achieved and each paper published, Wiles did not realise that he was gathering the experience which would many years later bring him to the verge of a proof for Fermat's Last Theorem.

Although nobody was aware of it at the time, the mathematicians of post-war Japan had already triggered a chain of events which would inextricably link elliptic equations to Fermat's Last Theorem. By encouraging Wiles to study elliptic equations, Coates had given him the tools which would later enable him to work on his dream.

Yutaka Taniyama

5

Proof by Contradiction

The mathematician's patterns, like the painter's or the poet's, must be beautiful; the ideas, like the colours or the words, must fit together in a harmonious way. Beauty is the first test: there is no permanent place in the world for ugly mathematics.

G.H. Hardy

In the January of 1954 a talented young mathematician at the University of Tokyo paid a routine visit to his departmental library. Goro Shimura was in search of a copy of *Mathematische Annalen*, Vol. 24. In particular he was after a paper by Deuring on his algebraic theory of complex multiplication, which he needed in order to help him with a particularly awkward and esoteric calculation.

To his surprise and dismay, the volume was already out. The borrower was Yutaka Taniyama, a vague acquaintance of Shimura who lived on the other side of the campus. Shimura wrote to Taniyama explaining that he urgently needed the journal to complete the nasty calculation, and politely asked when it would be returned.

A few days later, a postcard landed on Shimura's desk. Taniyama had replied, saying that he too was working on the exact same calculation and was stuck at the same point in the logic. He

Goro Shimura

suggested that they share their ideas and perhaps collaborate on the problem. This chance encounter over a library book ignited a partnership which would change the course of mathematical history.

Taniyama was born on 12 November 1927 in a small town a few miles north of Tokyo. The Japanese character symbolising his first name was intended to read 'Toyo', but most people outside his family misinterpreted it as 'Yutaka', and as Taniyama grew up he accepted and adopted this title. As a child Taniyama's education was constantly interrupted. He suffered several bouts of ill health, and during his teenage years he was struck down by tuberculosis and had to miss two years of high school. The onset of war caused even greater disruption to his schooling.

Goro Shimura, one year younger than Taniyama, had his education stopped altogether during the war years. His school was shut down and, instead of attending lessons, Shimura had to help the war effort by working in a factory assembling aircraft parts. Each evening he would attempt to make up for his lost schooling and in particular found himself drawn to mathematics. 'Of course there are many subjects to learn, but mathematics was the easiest because I could simply read mathematical textbooks. I learnt calculus by reading books. If I'd wanted to pursue chemistry or physics then I would have needed scientific equipment and I had no access to such things. I never thought that I was talented. I was just curious.'

A few years after the war had finished, Shimura and Taniyama found themselves at university. By the time they had exchanged postcards over the library book, life in Tokyo was beginning to return to normal and the two young academics could afford one or two small luxuries. They spent their afternoons in the coffee-shops, in the evenings they dined in a little restaurant specialising in whale meat, and at weekends they would stroll through the botanical gardens or the city park. All ideal locations for discussing their latest mathematical thoughts.

Although Shimura had a whimsical streak – even today he retains his fondness for Zen jokes – he was far more conservative and conventional than his intellectual partner. Shimura would rise at dawn and immediately get down to work, whereas his colleague would often still be awake at this time, having worked through the night. Visitors to his apartment would often find Taniyama fast asleep in the middle of the afternoon.

While Shimura was fastidious, Taniyama was sloppy to the point of laziness. Surprisingly this was a trait that Shimura admired: 'He was gifted with the special capability of making many mistakes, mostly in the right direction. I envied him for this and tried in vain to imitate him, but found it quite difficult to make good mistakes.'

Taniyama was the epitome of the absent-minded genius and this was reflected in his appearance. He was incapable of tying a decent knot, and so he decided that rather than tie his shoelaces a dozen times a day he would not tie them at all. He would always wear the same peculiar green suit with a strange metallic sheen. It was made from a fabric which was so outrageous that it had been rejected by the other members of his family.

When they met in 1954 Taniyama and Shimura were just beginning their mathematical careers. The tradition was, and still is, for young researchers to be taken under the wing of a professor who would guide the fledgling brain, but Taniyama and Shimura rejected this form of apprenticeship. During the war real research had ground to a halt and even by the 1950s the mathematics faculty had still not recovered. According to Shimura, the professors were 'tired, jaded and disillusioned'. In comparison the post-war students were passionate and eager to learn, and they soon realised that the only way forward would be for them to teach themselves. The students organised regular seminars, taking it in turn to inform each other of the latest techniques and breakthroughs. Despite his otherwise lackadaisical attitude, when it came to the seminars Taniyama provided a ferocious driving force. He would encourage the more senior students to explore uncharted territory, and for the younger students he acted as a father figure.

Because of their isolation, the seminars would occasionally cover subjects which were generally considered passé in Europe and America. The students' naïvety meant that they studied equations which had been abandoned in the West. One particularly unfashionable topic which fascinated both Taniyama and Shimura was the study of *modular forms*.

Modular forms are some of the weirdest and most wonderful objects in mathematics. They are one of the most esoteric entities in mathematics and yet the twentieth-century number theorist Martin Eichler rated them as one of the five fundamental operations: addition, subtraction, multiplication, division and modular forms. Most mathematicians would consider themselves masters of the first four operations, but the fifth one they still find a little confusing.

The key feature of modular forms is their inordinate level of symmetry. Although most people are familiar with the everyday concept of symmetry, it has a very particular meaning in mathematics, which is that an object has symmetry if it can be transformed in a particular way and yet afterwards appear to be unchanged. To appreciate the immense symmetry of a modular form it helps to first examine the symmetry of a more mundane object such as a simple square.

In the case of a square, one form of symmetry is rotational. That

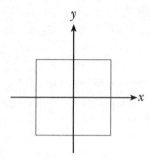

Figure 18. A simple square exhibits both rotational and reflectional symmetry.

is to say, if we imagine a pivot at the point where the x-axis and y-axis cross, then the square in Figure 18 can be rotated by one quarter of a turn, and afterwards it will appear to be unchanged. Similarly, rotations by half a turn, three-quarters of a turn and one full turn will also leave the square apparently unchanged.

In addition to rotational symmetry the square also possesses reflectional symmetry. If we imagine a mirror placed along the x-axis then the top half of the square would reflect exactly onto the lower half, and vice versa, so after the transformation the square would appear to remain unchanged. Similarly we can define three other mirrors (along the y-axis and along the two diagonals) for which the reflected square would appear to be identical to the original one.

The simple square is relatively symmetric, possessing both rotational and reflectional symmetries, but it does not possess any translational symmetry. This means that if the square were to be shifted in any direction, an observer would spot the movement immediately because its position relative to the axes would have

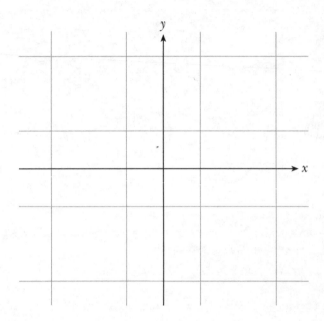

Figure 19. An infinite surface tiled with squares exhibits rotational and reflectional symmetry, and in addition has translational symmetry.

changed. However, if the whole of the space were tiled with squares, as shown in Figure 19, this infinite collection of squares would then have translational symmetry. If the infinite tiled surface were to be shifted up or down by one or more tile spaces, then the translated tiling would appear to be identical to the original one.

The symmetry of tiled surfaces is a relatively straightforward idea, but as with many seemingly simple concepts there are many

Figure 20. By using two different tiles, the kite and the dart, Roger Penrose was able to cover a surface. However, Penrose tiling does not possess translational symmetry.

subtleties hidden within it. For example, in the 1970s the British physicist and recreational mathematician Roger Penrose began dabbling with different tiles on the same surface. Eventually he identified two particularly interesting shapes, called the kite and the dart, which are shown in Figure 20. On their own each of these shapes could not be used to tile a surface without leaving gaps or overlaps, but together they could be used to create a rich set of tiling patterns. The kites and darts can be fitted together in an infinite number of ways, and although each pattern is apparently similar, in detail they all vary. One pattern made from kites and darts is shown in Figure 20.

Another remarkable feature of Penrose tilings (the patterns generated by tiles such as the kite and dart) is that they can exhibit a very restricted level of symmetry. At first sight it would appear that the tiling shown in Figure 20 would have translational symmetry, and yet any attempt to shift the pattern across so that it effectively remains unchanged ends in failure. Penrose tilings are deceptively unsymmetrical, and this is why they fascinate mathematicians and have become the starting point for a whole new area of mathematics.

Curiously Penrose tiling has also had repercussions in material science. Crystallographers always believed that crystals had to be built on the principles behind square tiling, possessing a high level of translational symmetry. In theory building crystals relied on a highly regular and repetitive structure. However, in 1984 scientists discovered a metallic crystal made of aluminium and manganese which was built along Penrose principles. The mosaic of aluminium and manganese behaved like the kites and darts, generating a crystal which was almost regular, but not quite. A French company has recently developed a Penrose crystal into a coating for frying-pans.

While the fascinating thing about Penrose's tiled surfaces is their restricted symmetry, the interesting property of modular forms is that they exhibit infinite symmetry. The modular forms studied by Taniyama and Shimura can be shifted, switched, swapped, reflected and rotated in an infinite number of ways and still they remain unchanged, making them the most symmetrical of mathematical objects. When the French polymath Henri Poincaré studied modular forms in the nineteenth century, he had great difficulty coming to terms with their immense symmetry. After working on a particular type of modular form, he described to his colleagues how every day for two weeks he would wake up and try and find an error in his calculations. On the fifteenth day he realised and accepted that modular forms were indeed symmetrical in the extreme.

Figure 21. Mauritz Escher's *Circle Limit IV* conveys some of the symmetry of modular forms.

Unfortunately, drawing, or even imagining, a modular form is impossible. In the case of the square tiling we have an object which lives in two dimensions, its space being defined by the *x*-axis and the y-axis. A modular form is also defined by two axes, but the axes are both complex, i.e. each axis has a real and an imaginary part and effectively becomes two axes. Therefore the first complex axis must be represented by two axes, x_r-axis (real) and x_i-axis (imaginary), and the second complex axis is represented by two axes, y_r-axis (real) and y_i-axis (imaginary). To be precise, modular forms live in the upper half-plane of this complex space, but what is most important to appreciate is that this is a four-dimensional space (x_r, x_i, y_r, y_i).

This four-dimensional space is called *hyperbolic space*. The hyperbolic universe is tricky to comprehend for humans, who are constrained to living in a conventional three-dimensional world, but four-dimensional space is a mathematically valid concept, and it is this extra dimension which gives the modular forms such an immensely high level of symmetry. The artist Mauritz Escher was fascinated by mathematical ideas and attempted to convey the concept of hyperbolic space in some of his etchings and paintings. Figure 21 shows Escher's *Circle Limit IV* which embeds the hyperbolic world into the two-dimensional page. In true hyperbolic space the bats and angels would be the same size, and the repetition is indicative of the high level of symmetry. Although some of this symmetry can be seen on the two-dimensional page, there is an increasing distortion towards the edge of the picture.

The modular forms which live in hyperbolic space come in various shapes and sizes, but each one is built from the same basic ingredients. What differentiates each modular form is the amount of each ingredient it contains. The ingredients of a modular form are labelled from one to infinity $(M_1, M_2, M_3, M_4, \ldots)$ and so a particular modular form might contain one lot of ingredient one $(M_1 = 1)$, three lots of ingredient two $(M_2 = 3)$, two lots of ingredient three $(M_3 = 2)$, etc. This information describing how a modular form is constructed can be summarised in a so-called modular series, or M-series, a list of the ingredients and the quantity of each one required:

$$M\text{-series:} \qquad M_1 = 1,$$
$$M_2 = 3,$$
$$M_3 = 2.$$
$$\vdots$$

Just as the E-series is the DNA for elliptic equations, the M-series is the DNA for modular forms. The amount of each ingredient listed in the M-series is critical. Depending how you change the amount of, say, the first ingredient you might generate a completely different, but equally symmetrical, modular form, or you might destroy the symmetry altogether and generate a new object which is not a modular form. If the quantity of each ingredient is arbitrarily chosen, then the result will probably be an object with little or no symmetry.

Modular forms stand very much on their own within mathematics. In particular, they would seem to be completely unrelated to the subject that Wiles would study at Cambridge, elliptic equations. The modular form is an enormously complicated beast, studied largely because of its symmetry and only discovered in the nineteenth century. The elliptic equation dates back to the ancient Greeks and has nothing to do with symmetry. Modular forms and elliptic equations live in completely different regions of the mathematical cosmos, and nobody would ever have believed that there was the remotest link between the two subjects. However, Taniyama and Shimura were to shock the mathematical community by suggesting that elliptic equations and modular forms were effectively one and the same thing. According to these two maverick mathematicians, they could unify the modular and elliptic worlds.

In 1955 Goro Shimura and Yutaka Taniyama attended an international symposium in Tokyo.

Wishful Thinking

In September 1955 an international symposium was held in Tokyo. It was a unique opportunity for the many young Japanese researchers to show off to the rest of the world what they had learned. They handed round a collection of thirty-six problems related to their work, accompanied by a humble introduction – *Some unsolved problems in mathematics: no mature preparation has been made, so there may be some trivial or already solved ones among these. The participants are requested to give comments on any of these problems.*

Four of the questions were from Taniyama, and these hinted at a curious relationship between modular forms and elliptic equations. These innocent questions would ultimately lead to a revolution in number theory. Taniyama had looked at the first few terms in the M-series of a particular modular form. He recognised the pattern and realised that it was identical to the list of numbers in the E-series of a well-known elliptic equation. He calculated a few more terms in each series, and still the M-series of the modular form and E-series of the elliptic equation matched perfectly.

This was an astonishing discovery because, for no apparent reason, this modular form could be related to an elliptic equation through their respective M-series and E-series – these series were identical. The mathematical DNA which made up these two entities was exactly the same. This was a doubly profound discovery. First, it suggested that deep down there was a fundamental relationship between the modular form and the elliptic equation, objects which come from opposite ends of mathematics. Second, it meant that mathematicians, who already knew the M-series for the modular form, would not have to calculate the E-series for the corresponding elliptic equation because it would be the same as the M-series.

Relationships between apparently different subjects are as creatively important in mathematics as they are in any discipline. The relationship hints at some underlying truth which enriches both subjects. For instance, originally scientists had studied electricity and magnetism as two completely separate phenomena. Then, in the nineteenth century, theorists and experimentalists realised that electricity and magnetism were intimately related. This resulted in a deeper understanding of both of them. Electric currents generate magnetic fields, and magnets can induce electricity in wires passing close to them. This led to the invention of dynamos and electric motors, and ultimately the discovery that light itself is the result of magnetic and electric fields oscillating in harmony.

Taniyama examined a few other modular forms and in each case the M-series seemed to correspond perfectly with the E-series of an elliptic equation. He began to wonder if it could be that every single modular form could be matched with an elliptic equation. Perhaps every modular form has the same DNA as an elliptic equation: perhaps each modular form is an elliptic equation in disguise? The questions he handed out at the symposium were related to this hypothesis.

The idea that every elliptic equation was related to a modular form was so extraordinary that those who glanced at Taniyama's questions treated them as nothing more than a curious observation. Sure enough Taniyama had demonstrated that a few elliptic equations could be related to particular modular forms, but they claimed that this was nothing more than a coincidence. According to the sceptics Taniyama's claim of a more general and universal relationship seemed to be largely unsubstantiated. The hypothesis was based on intuition rather than on any real evidence.

Taniyama's only ally was Shimura, who believed in the power and depth of his friend's idea. Following the symposium he worked with Taniyama in an attempt to develop the hypothesis to a level where the rest of the world could no longer ignore their work. Shimura wanted to find more evidence to back up the relationship between the modular and elliptic worlds. The collaboration was temporarily halted when in 1957 Shimura was invited to attend the Institute for Advanced Study in Princeton. Following his two years as a visiting professor in America he intended to resume working with Taniyama, but this was never to happen. On 17 November 1958, Yutaka Taniyama committed suicide.

Goro Shimura still has the last letter he received from his friend and colleague Yutaka Taniyama.

Death of a Genius

Shimura still keeps the postcard that Taniyama sent him when they first made contact over the library book. He also keeps the last letter Taniyama wrote to him while he was away in Princeton, but it contains not the merest hint as to what would happen just two months later. To this day Shimura has no understanding of what was behind Taniyama's suicide. 'I was very much puzzled. Puzzlement may be the best word. Of course I was sad, but it was so sudden. I got his letter in September and he died in early November, and I was unable to make sense out of this. Of course, later I heard various things and I tried to reconcile myself with his death. Some people said that he lost confidence in himself but not mathematically.'

What was particularly confusing for Taniyama's friends was that he had just fallen in love with Misako Suzuki and planned to marry her later that year. In a personal tribute published in the *Bulletin of the London Mathematical Society*, Goro Shimura recollects Taniyama's engagement to Misako and the weeks which led up to his suicide:

When informed of their engagement, I was somewhat surprised, since I had vaguely thought she was not his type, but I felt no misgivings. I was told afterward that they had signed a lease for an apartment, apparently a better one, for their new home, had bought some kitchenware together, and had been preparing for their wedding. Everything looked promising for them and their friends. Then the catastrophe befell them.

On the morning of Monday, November 17, 1958, the superintendent of his apartment found him dead in his room with a note left on a desk. It was written on three pages of a notebook of the type he had been using for his scholastic work; its first paragraph read like this:

'Until yesterday, I had no definite intention of killing myself. But more than a few must have noticed that lately I have been tired both physically and mentally. As to the cause of my suicide, I don't quite understand it myself, but it is not the result of a particular incident, nor of a specific matter. Merely may I say, I am in the frame of mind that I lost confidence in my future. There may be someone to whom my suicide will be troubling or a blow to a certain degree. I sincerely hope that this incident will cast no dark shadow over the future of that person. At any rate, I cannot deny that this is a kind of betrayal, but please excuse it as my last act in my own way, as I have been doing my own way all my life.'

He went on to describe, quite methodically, his wish of how his belongings should be disposed of, and which books and records were the ones he had borrowed from the library or from his friends, and so on. Specifically he says: 'I would like to leave the records and the player to Misako Suzuki provided she will not be upset by me leaving them to her'. Also he explains how far he reached in the undergraduate courses on calculus and linear algebra he was teaching, and concludes the note with an apology to his colleagues for the inconveniences this act could cause.

Thus one of the most brilliant and pioneering minds of the time ended his life by his own will. He had attained the age of thirty-one only five days earlier.

A few weeks after the suicide, tragedy struck a second time. His fiancée, Misako Suzuki, also took her own life. She reportedly left a note which read: 'We promised each other that no matter where we went, we would never be separated. Now that he is gone, I must go too in order to join him.'

Philosophy of Goodness

During his short career Taniyama contributed many radical ideas to mathematics. The questions he handed out at the symposium contained his greatest insight, but it was so ahead of its time that he would never live to see its enormous influence on number theory. His intellectual creativity was to be sadly missed, along with his guiding role within the community of young Japanese scientists. Shimura clearly remembers Taniyama's influence: 'He was always kind to his colleagues, especially to his juniors, and he genuinely cared about their welfare. He was the moral support of many of those who came into mathematical contact with him, including of course myself. Probably he was never concious of this role he was playing. But I feel his noble generosity in this respect even more strongly now than when he was alive. And yet nobody was able to give him any support when he desperately needed it. Reflecting on this, I am overwhelmed by the bitterest grief.'

Following Taniyama's death, Shimura concentrated all his efforts on understanding the exact relationship between elliptic equations and modular forms. As the years passed he struggled to gather more evidence and one or two pieces of logic to support the theory. Gradually he became increasingly convinced that every single elliptic equation must be related to a modular form. Other mathematicians were still dubious and Shimura recalls a conversation with an eminent colleague. The professor inquired, 'I hear that you propose that some elliptic equations can be linked to modular forms.'

'No, you don't understand,' replied Shimura. 'It's not just *some* elliptic equations, it's *every* elliptic equation!'

Shimura could not prove that this was the case but every time he tested the hypothesis it seemed to be true, and in any case it all seemed to fit in with his broad mathematical philosophy. 'I have this philosophy of goodness. Mathematics should contain goodness. So in the case of the elliptic equation, one might call the equation good if it is parametrised by a modular form. I expect all elliptic equations to be good. It's a rather crude philosophy but one can always take it as a starting point. Then, of course, I had to develop various technical reasons for the conjecture. I might say that the conjecture stemmed from that philosophy of goodness. Most mathematicians do mathematics from an aesthetic point of view and that philosophy of goodness comes from my aesthetic viewpoint.'

Eventually Shimura's accumulation of evidence meant that his theory about elliptic equations and modular forms became more widely accepted. He could not prove to the rest of the world that it was true, but at least it was now more than mere wishful thinking. There was enough evidence for it to be worthy of the title of conjecture. Initially it was referred to as the Taniyama–Shimura conjecture in recognition of the man who inspired it and his colleague who went on to develop it fully.

In due course André Weil, one of the godfathers of twentieth-century number theory, was to adopt the conjecture and publicise it in the West. Weil investigated the idea of Shimura and Taniyama, and found even more solid evidence in favour of it. As a result, the hypothesis was often referred to as the Taniyama–Shimura–Weil conjecture, sometimes as the Taniyama–Weil conjecture and occasionally as the Weil conjecture. In fact there has been much debate and controversy over the official naming of the conjecture. For those of you interested in combinatorics there are 15 possible permutations given the three names involved, and it is quite probable that every one of those combinations has appeared in print over the years. However, I will refer to the conjecture by its original title, the Taniyama–Shimura conjecture.

Professor John Coates, who guided Andrew Wiles when he was a student, was himself a student when the Taniyama–Shimura conjecture became a talking point in the West. 'I began research in 1966 when the conjecture of Taniyama and Shimura was sweeping through the world. Everyone was amazed and began to look seriously at the issue of whether all elliptic equations could be modular. This was a tremendously exciting time; the only problem, of course, was that it seemed very hard to make progress. I think it's fair to say that beautiful though this idea was it seemed very difficult to actually prove, and that's what we're primarily interested in as mathematicians.'

During the late sixties hoards of mathematicians repeatedly tested the Taniyama–Shimura conjecture. Starting with an elliptic equation and its E-series they would search for a modular form with an identical M-series. In every single case the elliptic equation did indeed have an associated modular form. Although this was good evidence in favour of the Taniyama–Shimura conjecture, it was by no means a proof. Mathematicians suspected that it was true, but until somebody could find a logical proof it would remain merely a conjecture.

Barry Mazur, a professor at Harvard University, witnessed the rise of the Taniyama–Shimura conjecture. 'It was a wonderful conjecture – the surmise that every elliptic equation is associated with a modular form – but to begin with it was ignored because it was so ahead of its time. When it was first proposed it was not taken up because it was so astounding. On the one hand you have the elliptic world, and on the other you have the modular world. Both these branches of mathematics had been studied intensively but separately. Mathematicians studying elliptic equations might not be well versed in things modular, and conversely. Then along comes the Taniyama–Shimura conjecture which is the grand surmise that there's a bridge between these two completely different worlds. Mathematicians love to build bridges.'

The value of mathematical bridges is enormous. They enable communities of mathematicians who have been living on separate islands to exchange ideas and explore each other's creations. Mathematics consists of islands of knowledge in a sea of ignorance. For example, there is the island occupied by geometers who study shape and form, and then there is the island of probability where mathematicians discuss risk and chance. There are dozens of such islands, each one with its own unique language, incomprehensible to the inhabitants of other islands. The language of geometry is quite different to the language of probability, and the slang of calculus is meaningless to those who speak only statistics.

The great potential of the Taniyama–Shimura conjecture was that it would connect two islands and allow them to speak to each other for the first time. Barry Mazur thinks of the Taniyama–Shimura conjecture as a translating device similar to the Rosetta stone, which contained Egyptian demotic, ancient Greek and hieroglyphics. Because demotic and Greek were already understood, archaeologists could decipher hieroglyphics for the first time. 'It's as if you know one language and this Rosetta stone is going to give you an intense understanding of the other language,' says Mazur. 'But the Taniyama–Shimura conjecture is a Rosetta stone with a certain magical power. The conjecture has the very pleasant feature that simple intuitions in the modular world translate into deep truths in the elliptic world, and conversely. What's more, very profound problems in the elliptic world can get solved sometimes by translating them using this Rosetta stone into the modular world, and discovering that we have the insights and tools in the modular world to treat the translated problem. Back in the elliptic world we would have been at a loss.'

If the Taniyama–Shimura conjecture was true it would enable mathematicians to tackle elliptic problems which had remained unsolved for centuries by approaching them through the modular world. The hope was that the fields of elliptic equations and modular forms could be unified. The conjecture also inspired the hope that links might exist between various other mathematical subjects.

During the 1960s Robert Langlands, at the Institute for Advanced Study, Princeton, was struck by the potency of the Taniyama–Shimura conjecture. Even though the conjecture had not been proved, Langlands believed that it was just one element of a much grander scheme of unification. He was confident that there were links between all the main mathematical topics and began to look for these unifications. Within a few years a number of links began to emerge. All these other unification conjectures were much weaker and more speculative than Taniyama–Shimura, but they formed an intricate network of hypothetical connections between many areas of mathematics. Langlands's dream was to see each of these conjectures proved one by one, leading to a grand unified mathematics.

Langlands discussed his plan for the future and tried to persuade other mathematicians to take part in what became known as the Langlands programme, a concerted effort to prove his myriad of conjectures. There seemed to be no obvious way to prove such speculative links, but if the dream could be made a reality then the reward would be enormous. Any insoluble problem in one area of mathematics could be transformed into an analogous problem in another area, where a whole new arsenal of techniques could be brought to bear on it. If a solution was still elusive, the problem could be transformed and transported to yet another area of mathematics, and so on, until it was solved. One day, according to the Langlands programme, mathematicians would be able to solve their most esoteric and intractable problems by shuffling them around the mathematical landscape.

There were also important implications for the applied sciences and engineering. Whether it is modelling the interactions between colliding quarks or discovering the most efficient way to organise a telecommunications network, often the key to the problem is performing a mathematical calculation. In some areas of science and technology the complexity of the calculations is so immense that progress in the subject has been severely hindered. If only mathematicians could prove the linking conjectures of the Langlands programme, then there would be short cuts to solving real-world problems, as well as abstract ones.

By the 1970s the Langlands programme had become a blueprint for the future of mathematics, but this route to a problem-solver's paradise was blocked by the simple fact that nobody had any real idea how to prove any of Langlands's conjectures. The strongest conjecture within the programme was still Taniyama–Shimura, but even this seemed out of reach. A proof of the Taniyama–Shimura conjecture would be the first step in the Langlands programme, and as such it had become one of the biggest prizes in modern number theory

Despite its status as a unproven conjecture, Taniyama–Shimura was still mentioned in hundreds of mathematical research papers speculating about what would happen if it could be proved. The papers would begin by clearly stating the caveat 'Assuming that the Taniyama–Shimura conjecture is true . . .', and then they would continue to outline a solution for some unsolved problem. Of course, these results could themselves only be hypothetical, because they relied on the Taniyama–Shimura conjecture being true. These new hypothetical results were in turn incorporated into other results until there existed a plethora of mathematics which relied on the truth of the Taniyama–Shimura conjecture. This one conjecture was a foundation for a whole new architecture of mathematics, but until it could be proved the whole structure was vulnerable.

At the time, Andrew Wiles was a young researcher at Cambridge University, and he recalls the trepidation that plagued the mathematics community in the 1970s: 'We built more and more conjectures which stretched further and further into the future, but they would all be ridiculous if the Taniyama–Shimura conjecture was not true. So we had to prove Taniyama–Shimura to show that this whole design we had hopefully mapped out for the future was correct.'

Mathematicians had constructed a fragile house of cards. They dreamed that one day someone would give their architecture the solid foundation it needed. They also had to live with the nightmare that one day someone might prove that Taniyama and Shimura were in fact wrong, causing two decades' worth of research to crash to the ground.

The Missing Link

During the autumn of 1984 a select group of number theorists gathered for a symposium in Oberwolfach, a small town in the heart of Germany's Black Forest. They had been brought together to discuss various breakthroughs in the study of elliptic equations, and naturally some of the speakers would occasionally report any minor progress that they had made towards proving the Taniyama–Shimura conjecture. One of the speakers, Gerhard Frey, a mathematician from Saarbrücken, could not offer any new ideas as to how to attack the conjecture, but he did make the remarkable claim that if anyone could prove the Taniyama–Shimura conjecture then they would also immediately prove Fermat's Last Theorem.

When Frey got up to speak he began by writing down Fermat's equation:

$$x^n + y^n = z^n \quad \text{where } n \text{ is greater than 2.}$$

Fermat's Last Theorem claims that there are no whole number solutions to this equation, but Frey explored what would happen if the Last Theorem were false, i.e. that there is at least one solution. Frey had no idea what his hypothetical, and heretical, solution might be and so he labelled the unknown numbers with the letters A, B and C:

$$A^N + B^N = C^N.$$

Frey then proceeded to 'rearrange' the equation. This is a rigorous mathematical procedure which changes the appearance of the equation without altering its integrity. By a deft series of complicated manoeuvres Frey fashioned Fermat's original equation, with the hypothetical solution, into

$$y^2 = x^3 + (A^N - B^N)x^2 - A^N B^N.$$

Although this rearrangement seems very different from the original equation, it is a direct consequence of the hypothetical solution. That is to say if, and it is a big 'if', there is a solution to Fermat's equation and Fermat's Last Theorem is false, then this rearranged equation must also exist. Initially Frey's audience was not particularly impressed by his rearrangement, but then he pointed out that this new equation was in fact an elliptic equation, albeit a rather convoluted and exotic one. Elliptic equations have the form

$$y^2 = x^3 + ax^2 + bx + c,$$

but if we let

$$a = A^N - B^N, \qquad b = 0, \qquad c = -A^N B^N,$$

then it is easier to appreciate the elliptical nature of Frey's equation.

By turning Fermat's equation into an elliptic equation, Frey had linked Fermat's Last Theorem to the Taniyama–Shimura conjecture. Frey then pointed out to his audience that his elliptic equation, created from the solution to the Fermat equation, is truly bizarre. In fact, Frey claimed that his elliptic equation is so weird that the repercussions of its existence would be devastating for the Taniyama–Shimura conjecture.

Remember that Frey's elliptic equation is only a phantom equation. Its existence is conditional on that fact that Fermat's Last Theorem is false. However, if Frey's elliptic equation does exist, then it is so strange that it would be seemingly impossible for it ever to be related to a modular form. But the Taniyama–Shimura conjecture claims that *every* elliptic equation must be related to a modular form. Therefore the existence of Frey's elliptic equation defies the Taniyama–Shimura conjecture.

In other words, Frey's argument was as follows:

(1) If (and only if) Fermat's Last Theorem is wrong, then Frey's elliptic equation exists.
(2) Frey's elliptic equation is so weird that it can never be modular.
(3) The Taniyama–Shimura conjecture claims that every elliptic equation must be modular.
(4) Therefore the Taniyama–Shimura conjecture must be false!

Alternatively, and more importantly, Frey could run his argument backwards:

(1) If the Taniyama–Shimura conjecture can be proved to be true, then every elliptic equation must be modular.
(2) If every elliptic equation must be modular, then the Frey elliptic equation is forbidden to exist.
(3) If the Frey elliptic equation does not exist, then there can be no solutions to Fermat's equation.
(4) Therefore Fermat's Last Theorem is true!

Gerhard Frey had come to the dramatic conclusion that the truth of Fermat's Last Theorem would be an immediate consequence of the Taniyama–Shimura conjecture being proved. Frey claimed that if mathematicians could prove the Taniyama–Shimura conjecture then they would automatically prove Fermat's Last Theorem. For the first time in a hundred years the world's hardest mathematical problem looked vulnerable. According to Frey, proving the Taniyama–Shimura conjecture was the only hurdle to proving Fermat's Last Theorem.

Although the audience was impressed by Frey's brilliant insight, they were also struck by an elementary blunder in his logic. Almost everyone in the auditorium, except Frey himself, had spotted it. The mistake did not appear to be serious: nonetheless as it stood Frey's work was incomplete. Whoever could correct the error first would take the credit for linking Fermat and Taniyama–Shimura.

Frey's audience dashed out of the lecture theatre and headed for the photocopying room. Often the importance of a talk can be gauged by the length of the queue waiting to run off copies of the lecture. Once they had a complete outline of Frey's ideas, they returned to their respective institutes and began to try and fill in the gap.

Frey's argument depended on the fact that his elliptic equation derived from Fermat's equation was so weird that it was not modular. His work was incomplete because he had not quite demonstrated that his elliptic equation was sufficiently weird. Only when somebody could prove the *absolute* weirdness of Frey's elliptic equation would a proof of the Taniyama–Shimura conjecture then imply a proof of Fermat's Last Theorem.

Initially mathematicians believed that proving the weirdness of Frey's elliptic equation would be a fairly routine process. At first sight Frey's mistake seemed to have been elementary and everyone who had been at Oberwolfach assumed that it was going to be a race to see who could the shuffle the algebra most quickly. The expectation was that somebody would send out an e-mail within a matter of days describing how they had established the true weirdness of Frey's elliptic equation.

A week passed and there was no such e-mail. Months passed and what was supposed to be a mathematical mad dash was turning into a marathon. It seemed that Fermat was still teasing and tormenting his descendants. Frey had outlined a tantalising strategy for proving Fermat's Last Theorem, but even the first elementary step, proving that Frey's hypothetical elliptic equation was not modular, was baffling mathematicians around the globe.

To prove that an elliptic equation is not modular, mathematicians were looking for invariants similar to those described in Chapter 4. The knot invariant showed that one knot could not be transformed into another, and Loyd's puzzle invariant showed that his 14–15 puzzle could not be transformed into the correct arrangement. If number theorists could discover an appropriate invariant to describe Frey's elliptic equation, then they could prove that, no matter what was done to it, it could never be transformed into a modular form.

One of those toiling to prove and complete the connection

Ken Ribet

between the Taniyama–Shimura conjecture and Fermat's Last Theorem was Ken Ribet, a professor at the University of California at Berkeley. Since the lecture at Oberwolfach, Ribet had become obsessed with trying to prove that Frey's elliptic equation was too weird to be modular. After eighteen months of effort he, along with everybody else, was getting nowhere. Then, in the summer of 1986, Ribet's colleague Professor Barry Mazur was visiting Berkeley to attend the International Congress of Mathematicians. The two friends met up for a cappuccino at the Café Strada and began sharing bad luck stories and grumbling about the state of mathematics.

Eventually they started discussing the latest news on the various attempts to prove the weirdness of Frey's elliptic equation, and Ribet began explaining a tentative strategy which he had been exploring. The approach seemed vaguely promising but he could only prove a very minor part of it. 'I sat down with Barry and told him what I was working on. I mentioned that I'd proved a very special case, but I didn't know what to do next to generalise it to get the full strength of the proof.'

Professor Mazur sipped his cappuccino and listened to Ribet's idea. Then he stopped and stared at Ken in disbelief. 'But don't you see? You've already done it! All you have to do is add some gamma-zero of (M) structure and just run through your argument and it works. It gives you everything you need.'

Ribet looked at Mazur, looked at his cappuccino, and looked back at Mazur. It was the most important moment of Ribet's career and he recalls it in loving detail. 'I said you're absolutely right – of course – how did I not see this? I was completely astonished because it had never occurred to me to add the extra gamma-zero of (M) structure, simple as it sounds.'

It should be noted that, although *adding gamma-zero of (M) structure* sounds simple to Ken Ribet, it is an esoteric step of logic which only a handful of the world's mathematicians could have concocted over a casual cappuccino.

'It was the crucial ingredient that I had been missing and it had been staring me in the face. I wandered back to my apartment on a cloud, thinking: My God is this really correct? I was completely enthralled and I sat down and started scribbling on a pad of paper. After an hour or two I'd written everything out and verified that I knew the key steps and that it all fitted together. I ran through my argument and I said, yes, this absolutely has to work. And there were of course thousands of mathematicians at the International Congress and I sort of casually mentioned to a few people that I'd proved that the Taniyama–Shimura conjecture implies Fermat's Last Theorem. It spread like wildfire and soon large groups of people knew; they were running up to me asking, *Is it really true you've proved that Frey's elliptic equation is not modular?* And I had to think for a minute and all of a sudden I said, *Yes, I have.*'

Fermat's Last Theorem was now inextricably linked to the Taniyama–Shimura conjecture. If somebody could prove that every elliptic equation is modular, then this would imply that Fermat's equation had no solutions, and immediately prove Fermat's Last Theorem.

For three and half centuries Fermat's Last Theorem had been an isolated problem, a curious and impossible riddle on the edge of mathematics. Now Ken Ribet, inspired by Gerhard Frey, had brought it centre stage. The most important problem from the seventeenth century was coupled to the most significant problem of the twentieth century. A puzzle of enormous historical and emotional importance was linked to a conjecture that could revolutionise modern mathematics. In effect, mathematicians could now attack Fermat's Last Theorem by adopting a strategy of proof by contradiction. To prove that the Last Theorem is true, mathematicians would begin by assuming it to be false. The implication of being false would be to make the Taniyama–Shimura conjecture false. However, if Taniyama–Shimura could be proven true, then this would be incompatible with Fermat's Last Theorem being false, therefore it, too, would have to be true.

Frey had clearly defined the task ahead. Mathematicians would automatically prove Fermat's Last Theorem if they could first prove the Taniyama–Shimura conjecture.

Initially there was renewed hope but then the reality of the situation dawned. Mathematicians had been trying to prove Taniyama–Shimura for thirty years and they had failed. Why should they make any progress now? The sceptics believed that what little hope there was of proving the Taniyama–Shimura conjecture had now vanished. Their logic was that anything that might lead to a solution of Fermat's Last Theorem must, by definition, be impossible.

Even Ken Ribet, who had made the crucial breakthrough, was pessimistic: 'I was one of the vast majority of people who believed that the Taniyama–Shimura conjecture was completely inaccessible. I didn't bother to try and prove it. I didn't even think about trying to prove it. Andrew Wiles was probably one of the few people on earth who had the audacity to dream that you can actually go and prove this conjecture.'

In 1986 Andrew Wiles realised that it might be possible to prove Fermat's Last Theorem via the Taniyama–Shimura conjecture.

6

The Secret Calculation

An expert problem solver must be endowed with two incompatible qualities – a restless imagination and a patient pertinacity.

Howard W. Eves

'It was one evening at the end of the summer of 1986 when I was sipping iced tea at the house of a friend. Casually in the middle of a conversation he told me that Ken Ribet had proved the link between Taniyama–Shimura and Fermat's Last Theorem. I was electrified. I knew that moment that the course of my life was changing because this meant that to prove Fermat's Last Theorem all I had to do was to prove the Taniyama–Shimura conjecture. It meant that my childhood dream was now a respectable thing to work on. I just knew that I could never let that go. I just knew that I would go home and work on the Taniyama–Shimura conjecture.'

Over two decades had passed since Andrew Wiles had discovered the library book that inspired him to take up Fermat's challenge, but now, for the first time, he could see a path towards achieving his childhood dream. Wiles recalls how his attitude to Taniyama–Shimura changed overnight: 'I remembered one mathematician who'd written about the Taniyama–Shimura conjecture and cheekily suggested it as an exercise for the interested reader. Well, I guess now I was interested!'

Since completing his Ph.D. with Professor John Coates at Cambridge, Wiles had moved across the Atlantic to Princeton University where he himself was now a professor. Thanks to Coates's guidance Wiles probably knew more about elliptic equations than anybody else in the world, but he was well aware that even with his enormous background knowledge and mathematical skills the task ahead was immense.

Most other mathematicians, including John Coates, believed that embarking on the proof was a futile exercise: 'I myself was very sceptical that the beautiful link between Fermat's Last Theorem and the Taniyama–Shimura conjecture would actually lead to anything, because I must confess I did not think that the Taniyama–Shimura conjecture was accessible to proof. Beautiful though this problem was, it seemed impossible to actually prove. I must confess I thought I probably wouldn't see it proved in my lifetime.'

Wiles was aware that the odds were against him, but even if he ultimately failed in proving Fermat's Last Theorem he felt his efforts would not be wasted: 'Of course the Taniyama–Shimura conjecture had been open for many years. No one had had any idea how to approach it but at least it was mainstream mathematics. I could try and prove results, which, even if they didn't get the whole thing, would be worthwhile mathematics. I didn't feel I'd be wasting my time. So the romance of Fermat which had held me all my life was now combined with a problem that was professionally acceptable.'

The Attic Recluse

At the turn of the century the great logician David Hilbert was asked why he never attempted a proof of Fermat's Last Theorem.

He replied, 'Before beginning I should have to put in three years of intensive study, and I haven't that much time to squander on a probable failure.' Wiles realised that to have any hope of finding a proof he would first have to completely immerse himself in the problem, but unlike Hilbert he was prepared to take the risk. He read all the most recent journals and then played with the latest techniques over and over again until they became second nature to him. Gathering the necessary weapons for the battle ahead would require Wiles to spend the next eighteen months familiarising himself with every bit of mathematics which had ever been applied to, or had been derived from, elliptic equations or modular forms. This was a comparatively minor investment, bearing in mind that he fully expected that any serious attempt on the proof could easily require ten years of single-minded effort.

Wiles abandoned any work which was not directly relevant to proving Fermat's Last Theorem and stopped attending the never-ending round of conferences and colloquia. Because he still had responsibilities in the Princeton Mathematics Department, Wiles continued to attend seminars, lecture to undergraduates and give tutorials. Whenever possible he would avoid the distractions of being a faculty member by working at home where he could retreat into his attic study. Here he would attempt to expand and extend the power of the established techniques, hoping to develop a strategy for his attack on the Taniyama–Shimura conjecture.

'I used to come up to my study, and start trying to find patterns. I tried doing calculations which explain some little piece of mathematics. I tried to fit it in with some previous broad conceptual understanding of some part of mathematics that would clarify the particular problem I was thinking about. Sometimes that would involve going and looking it up in a book to see how it's done there. Sometimes it was a question of modifying things a bit, doing a little extra calculation. And sometimes I realised that nothing that had ever been done before was any use at all. Then I just had to find something completely new – it's a mystery where that comes from.

'Basically it's just a matter of thinking. Often you write something down to clarify your thoughts, but not necessarily. In particular when you've reached a real impasse, when there's a real problem that you want to overcome, then the routine kind of mathematical thinking is of no use to you. Leading up to that kind of new idea there has to be a long period of tremendous focus on the problem without any distraction. You have to really think about nothing but that problem – just concentrate on it. Then you stop. Afterwards there seems to be a kind of period of relaxation during which the subconscious appears to take over and it's during that time that some new insight comes.'

From the moment he embarked on the proof, Wiles made the remarkable decision to work in complete isolation and secrecy. Modern mathematics has developed a culture of cooperation and collaboration, and so Wiles's decision appeared to hark back to a previous era. It was as if he was imitating the approach of Fermat himself, the most famous of mathematical hermits. Wiles explained that part of the reason for his decision to work in secrecy was his desire to work without being distracted: 'I realised that anything to do with Fermat's Last Theorem generates too much interest. You can't really focus yourself for years unless you have undivided concentration, which too many spectators would have destroyed.'

Another motivation for Wiles's secrecy must have been his craving for glory. He feared the situation arising whereby he had completed the bulk of the proof but was still missing the final element of the calculation. At this point, if news of his breakthroughs were to leak out, there would be nothing stopping a rival mathematician building on Wiles's work, completing the proof and stealing the prize.

In the years to come Wiles was to make a series of extraordinary discoveries, none of which would be discussed or published until his proof was complete. Even close colleagues were oblivious to his research. John Coates can recall exchanges with Wiles during which he was given no clues as to what was going on: 'I remember saying to him on a number of occasions, "It's all very well this link to Fermat's Last Theorem but it's still hopeless to try and prove Taniyama–Shimura." I think he just smiled.'

Ken Ribet, who completed the link between Fermat and Taniyama–Shimura, was also completely unaware of Wiles's clandestine activities. 'This is probably the only case I know where someone worked for such a long time without divulging what he was doing, without talking about the progress he was making. It's just unprecedented in my experience. In our community people have always shared their ideas. Mathematicians come together at conferences, they visit each other to give seminars, they send e-mail to each other, they talk on the telephone, they ask for insights, they ask for feedback – mathematicians are always in communication. When you talk to other people you get a pat on the back; people tell you that what you've done is important, they give you ideas. It's sort of nourishing and if you cut yourself off from this, then you are doing something that's probably psychologically very odd.'

In order not to arouse suspicion Wiles devised a cunning ploy which would throw his colleagues off the scent. During the early 1980s he had been working on a major piece of research on a particular type of elliptic equation, which he was about to publish in its entirety, until the discoveries of Ribet and Frey made him change his mind. Wiles decided to publish his research bit by bit, releasing another minor paper every six months or so. This apparent productivity would convince his colleagues that he was still continuing with his usual research. For as long as he could maintain this charade, Wiles could continue working on his true obsession without revealing any of his breakthroughs.

The only person who was aware of Wiles's secret was his wife, Nada. They married soon after Wiles began working on the proof, and as the calculation progressed he confided in her and her alone. In the years that followed, his family would be his only distraction. 'My wife's only known me while I've been working on Fermat. I told her on our honeymoon, just a few days after we got married. My wife had heard of Fermat's Last Theorem, but at that time she had no idea of the romantic significance it had for mathematicians, that it had been such a thorn in our flesh for so many years.'

Duelling with Infinity

In order to prove Fermat's Last Theorem Wiles had to prove the Taniyama–Shimura conjecture: every single elliptic equation can be correlated with a modular form. Even before the link to Fermat's Last Theorem mathematicians had tried desperately to prove the conjecture, but every attempt had ended in failure. Wiles was acquainted with the failures of the past: 'Ultimately what one would naïvely have tried to do, and what people certainly did try to do, was to count elliptic equations and count modular forms, and show that there are the same number of each. But nobody has ever found any simple way of doing that. The first problem is that there are an infinite number of each and you can't count an infinite number. One simply doesn't have a way of doing it.'

In order to find a solution, Wiles adopted his usual approach to solving difficult problems. 'I sometimes write scribbles or doodles. They're not important doodles, just subconscious doodles. I never use a computer.' In this case, as with many problems in number theory, computers would be of no use whatsoever. The Tani-yama–Shimura conjecture applied to an infinite number of equa-tions and, although a computer could check an individual case in a few seconds, it could never check all cases. Instead what was required was a logical step-by-step argument which would effec-tively give a reason and explain why every elliptic equation had to be modular. To find the proof Wiles relied solely on a piece of paper, a pencil and his mind. 'I carried this thought around in my head basically the whole time. I would wake up with it first thing in the morning, I would be thinking about it all day and I would be thinking about it when I went to sleep. Without distraction I would have the same thing going round and round in my mind.'

After a year of contemplation Wiles decided to adopt a general strategy known as *induction* as the basis for his proof. Induction is an immensely powerful form of proof, because it can allow a math-ematician to prove that a statement is true for an infinite number of cases by only proving it for just one case. For example, imagine that a mathematician wants to prove that a statement is true for every counting number up to infinity. The first step is to prove that the statement is true for the number 1, which presumably is a fairly straightforward task. The next step is to show that if the statment is true for the number 1 then it must be true for the number 2, and if it is true for the number 2 then it must be true for the number 3, and if it is true for the number 3 then it must be true for the number 4, and so on. More generally, the mathematician has to show that if the statement is true for any number n, then it must be true for the next number $n + 1$.

Proof by induction is essentially a two step process:

(1) Prove that the statement is true for the first case.
(2) Prove that if the statement is true for any one case, then it must be true for the next case.

Another way to think of proof by induction is to imagine the infinite number of cases as an infinite line of dominoes. In order to prove every case it is necessary to find a way of knocking down every one of the dominoes. Knocking them down one by one would take an infinite amount of time and effort, but proof by induction allows mathematicians to knock them all down by just knocking down the first one. If the dominoes are carefully arranged, then knocking down the first domino will knock down the second domino, which will in turn knock down the third domino, and so on to infinity. Proof by induction invokes the domino effect. This form of mathematical domino-toppling allows an infinite number of cases to be proved by just proving the first one. Appendix 10 shows how proof by induction can be used to prove a relatively simple mathematical statement about all numbers.

The challenge for Wiles was to construct an inductive argument which showed that each of the infinity of elliptic equations could be matched to each of the infinity of modular forms. Somehow he had to break the proof down into an infinite number of individual cases and then prove the first case. Next, he had to demonstrate that, having proved the first case, all the others would topple. Eventually he discovered the first step to his inductive proof hidden in the work of a tragic genius from nineteenth-century France.

Evariste Galois was born in Bourg-la-Reine, a small village just south of Paris, on 25 October 1811, just twenty-two years after the French Revolution. Napoleon Bonaparte was at the height of his

Evariste Galois

powers, but the following year saw the disastrous Russian campaign, and in 1814 he was driven into exile and replaced by King Louis XVIII. In 1815 Napoleon escaped from Elba, entered Paris and reclaimed power but within a hundred days he was defeated at Waterloo and forced to abdicate once again in favour of Louis XVIII. Galois, like Sophie Germain, grew up during a period of immense upheaval, but whereas Germain shut herself away from the turmoils of the French Revolution and concentrated on mathematics, Galois repeatedly found himself at the centre of political controversy, which not only distracted him from a brilliant academic career, but also led to his untimely death.

In addition to the general unrest which impinged on everybody's life, Galois's interest in politics was inspired by his father, Nicolas-Gabriel Galois. When Evariste was just four years old his father was elected mayor of Bourg-la-Reine. This was during Napoleon's triumphant return to power, a period when his father's strong liberal values were in keeping with the mood of the nation. Nicolas-Gabriel Galois was a cultured and gracious man and during his early years as mayor he gained respect throughout the community, so even when Louis XVIII returned to the throne he retained his elected position. Outside of politics, his main interest seems to have been the composition of witty rhymes, which he would read at town meetings to the delight of his constituents. Many years later this charming talent for epigrams would lead to his downfall.

At the age of twelve Evariste Galois attended his first school, the Lycée of Louis-le-Grand, a prestigious but authoritarian institution. To begin with he did not encounter any courses in mathematics and his academic record was respectable but not outstanding. However, one event occurred during his first term which would influence the course of his life. The Lycée had previously been a Jesuit school and rumours began to circulate suggesting that it was about to be returned to the authority of the priests. During this period there was a continual struggle between republicans and monarchists to sway the balance of power between Louis XVIII and the people's representatives, and the increasing influence of the priests was seen as an indication of a shift away from the people and towards the King. The students of the Lycée, who in the main had republican sympathies, planned a rebellion but the director of the school, Monsieur Berthod, uncovered the plot and immediately expelled the dozen or so ring-leaders. The following day when Berthod demanded a demonstration of allegiance from the remaining senior scholars, they refused to drink a toast to Louis XVIII, whereupon another hundred students were expelled. Galois was too young to be involved in the failed rebellion and so remained at the Lycée. Nevertheless, watching his fellow students being humiliated in this way only served to inflame his republican tendencies.

It was not until the age of sixteen that Galois enrolled in his first mathematics class, a course which would, in the eyes of his teachers, transform him from a conscientious pupil into an unruly student. His school reports show that he neglected all his other subjects and concentrated solely on his new found passion:

This student works only in the highest realms of mathematics. The mathematical madness dominates this boy. I think it would be best for him if his parents would allow him to study nothing but this. Otherwise he is wasting his time here and does nothing but torment his teachers and overwhelm himself with punishments.

Galois's desire for mathematics soon outstripped the capacity of his teacher, and so he learnt directly from the very latest books written by the masters of the age. He readily absorbed the most complex of concepts, and by the time he was seventeen he published his first paper in the *Annales de Gergonne*. The path ahead seemed clear for the prodigy, except that his own sheer brilliance was to provide the greatest obstacle to his progress. Although he obviously knew more than enough mathematics to pass the Lycée's examinations, Galois's solutions were often so innovative and sophisticated that his examiners failed to appreciate them. To make matters worse Galois would perform so many calculations in his head that he would not bother to outline clearly his argument on paper, leaving the inadequate examiners even more perplexed and frustrated.

The young genius did not help the situation by having a quick temper and a rashness which did not endear him to his tutors or anybody else who crossed his path. When Galois applied to the Ecole Polytechnique, the most prestigious college in the land, his abruptness and lack of explanation in the oral examination meant that he was refused admission. Galois was desperate to attend the Polytechnique, not just because of its academic excellence but also because of its reputation for being a centre for republican activism. One year later he reapplied and once again his logical leaps in the oral examination only served to confuse his examiner, Monsieur Dinet. Sensing that he was about to be failed for a second time and frustrated that his brilliance was not being recognised, Galois lost his temper and threw a blackboard rubber at Dinet, scoring a direct hit. Galois was never to return to the hallowed halls of the Polytechnique.

Undaunted by the rejections, Galois remained confident of his mathematical talent and continued his own private researches. His main interest concerned finding solutions to equations, such as quadratic equations. Quadratic equations have the form

$$ax^2 + bx + c = 0, \quad \text{where } a, b, c \text{ can have any value.}$$

The challenge is to find the values of x for which the quadratic equation holds true. Rather than relying on trial and error mathematicians would prefer to have a recipe for finding solutions, and fortunately such a recipe exists:

$$x = \frac{-b \pm \sqrt{(b^2 - 4ac)}}{2a}.$$

Simply by substituting the values for a, b and c into the above recipe one can calculate the correct values for x. For instance, we can apply the recipe to solve the following equation:

$$2x^2 - 6x + 4 = 0, \quad \text{where } a = 2, b = -6 \text{ and } c = 4.$$

By putting the values of a, b and c into the recipe, the solution turns out to be $x = 1$ or $x = 2$.

The quadratic is a type of equation within a much larger class of equations known as polynomials. A more complicated type of polynomial is the cubic equation:

$$ax^3 + bx^2 + cx + d = 0.$$

The extra complication comes from the additional term x^3. By adding one more term x^4, we get the next level of polynomial equation, known as the quartic:

$$ax^4 + bx^3 + cx^2 + dx + e = 0.$$

By the nineteenth century, mathematicians also had recipes which could be used to find solutions to the cubic and the quartic equations, but there was no known method for finding solutions to the quintic equation:

$$ax^5 + bx^4 + cx^3 + dx^2 + ex + f = 0.$$

Galois became obsessed with finding a recipe for solving quintic equations, one of the great challenges of the era, and by the age of seventeen he had made sufficient progress to submit two research papers to the Academy of Sciences. The referee appointed to judge the papers was Augustin-Louis Cauchy, who many years later would argue with Lamé over an ultimately flawed proof of Fermat's Last Theorem. Cauchy was highly impressed by the young man's work and judged it worthy of being entered for the Academy's Grand Prize in Mathematics. In order to qualify for the competition the two papers would have to be re-submitted in the form of a single memoir, so Cauchy returned them to Galois and awaited his entry.

Having survived the criticisms of his teachers and rejection by the Ecole Polytechnique Galois's genius was on the verge of being recognised, but over the course of the next three years a series of personal and professional tragedies would destroy his ambitions. In July of 1829 a new Jesuit priest arrived in the village of Bourg-le-Reine, where Galois's father was still mayor. The priest took exception to the mayor's republican sympathies and began a campaign to oust him from office by spreading rumours aimed at discrediting him. In particular the scheming priest exploited Nicolas-Gabriel Galois's reputation for composing clever rhymes. He wrote a series of vulgar verses ridiculing members of the community and signed them with the mayor's name. The elder Galois could not survive the shame and the embarrassment which resulted and decided that the only honourable option was to commit suicide.

Evariste Galois returned to attend his father's funeral and saw for himself the divisions that the priest had created in the village. As the coffin was being lowered into the grave, a scuffle broke out between the Jesuit priest, who was conducting the service, and supporters of the mayor, who realised that there had been a plot to undermine him. The priest suffered a gash to the head, the scuffle turned into a riot, and the coffin was left to drop unceremoniously into its grave. Watching the French establishment humiliate and destroy his father served only to consolidate Galois's fervent support for the republican cause.

Upon returning to Paris, Galois combined his research papers well ahead of the competition deadline and submitted the memoir to the secrerary of the Academy, Joseph Fourier, who was supposed to pass it on to the judging committee. Galois's paper did not offer a solution to the quintic problem but it did offer a brilliant insight and many mathematicians, including Cauchy, considered that it was a likely winner. To the shock of Galois and his friends, not only did he fail to win the prize, but he had not even been officially entered. Fourier had died a few weeks prior to the judging and, although a stack of competition entries was passed on to the committee, Galois's memoir was not among them. The memoir was never found and the injustice was recorded by a French journalist.

Last year before March 1st, Monsieur Galois gave to the secretary of the Institute a memoir on the solution of numerical equations. This memoir should have been entered in the competition for the Grand Prize in Mathematics. It deserved the prize, for it could resolve some difficulties that Lagrange had failed to do. Monsieur Cauchy had conferred the highest praise on the author about this subject. And what happened? The memoir is lost and the prize is given without the participation of the young savant.

Le Globe, 1831

Galois felt that his memoir had been deliberately lost by a politically biased Academy, a belief that was reinforced a year later when the Academy rejected his next manuscript, claiming that 'his argument is neither sufficiently clear nor sufficiently developed to allow us to judge its rigour'. He decided that there was a conspiracy to exclude him from the mathematical community, and as a result he neglected his research in favour of fighting for the republican cause. By this time he was a student at the Ecole Normale Supérieure, a slightly less prestigious college than the Ecole Polytechnique. At the Ecole Normale Galois's notoriety as a trouble-maker was overtaking his reputation as a mathematician. This culminated during the July revolution of 1830 when Charles X fled France and the political factions fought for control in the streets of Paris. The Ecole's director Monsieur Guigniault, a monarchist, was aware that the majority of his students were radical republicans and so confined them to their dormitories and locked the gates of the college. Galois was being prevented from fighting alongside his brothers, and his frustration and anger were compounded when the republicans were eventually defeated. When the opportunity arose he published a scathing attack on the college director, accusing him of cowardice. Not surprisingly, Guigniault expelled the insubordinate student and Galois's formal mathematical career was at an end.

On 4 December the thwarted genius attempted to become a professional rebel by joining the Artillery of the National Guard, a republican branch of the militia otherwise known as the 'Friends of the People'. Before the end of the month the new king Louis-Phillipe, anxious to avoid a further rebellion, abolished the Artillery of the National Guard, and Galois was left destitute and homeless. The most brilliant young talent in all of Paris was being persecuted at every turn and some of his former mathematical colleagues were becoming increasingly worried about his plight. Sophie Germain, who was by this time the shy elder stateswoman of French mathematics, expressed her concerns to friend of the family Count Libri-Carrucci:

Decidedly there is a misfortune concerning all that touches upon mathematics. The death of Monsieur Fourier has been the final blow for this student Galois who, in spite of his impertinence, showed signs of a clever disposition. He has been expelled from the Ecole Normale, he is without money, his mother has very little also and he continues his habit of insult. They say he will go completely mad. I fear this is true.

As long as Galois's passion for politics continued it was inevitable that his fortunes would deteriorate further, a fact documented by the great French writer Alexandre Dumas. Dumas was at the restaurant *Vendanges de Bourgogne* when he happened upon a celebration banquet in honour of nineteen republicans aquitted of conspiracy charges:

Suddenly, in the midst of a private conversation which I was carrying on with the person on my left, the name Louis-Phillipe, followed by five or six whistles, caught my ear. I turned around. One of the most animated scenes was taking place fifteen or twenty seats from me. It would be difficult to find in all Paris two hundred persons more hostile to the government than those to be found reunited at five o'clock in the afternoon in the long hall on the ground floor above the garden.

A young man who had raised his glass and held an open dagger in the same hand was trying to make himself heard – Evariste Galois was one of the most ardent republicans. The noise was such that the very reason for this noise had become incomprehensible. All that I could perceive was that there was a threat and that the name of Louis-Phillipe had been mentioned: the intention was made clear by the open knife.

This went way beyond my own republican opinions. I yielded to the pressure from my neighbour on the left who, as one of the King's comedians, didn't care to be compromised, and we jumped from the window sill into the garden. I went home somewhat worried. It was clear this episode would have its consequences. Indeed, two or three days later, Evariste Galois was arrested.

After being detained at Sainte-Pélagie prison for a month Galois was charged with threatening the King's life and brought to trial. Although there was little doubt from his actions that Galois was guilty, the raucous nature of the banquet meant that nobody could actually confirm that they had heard him make any direct threats. A sympathetic jury and the rebel's tender age – he was still only twenty – led to his acquittal. The following month he was arrested again.

On Bastille Day, 14 July 1831, Galois marched through Paris dressed in the uniform of the outlawed Artillery Guard. Although this was merely a gesture of defiance, he was sentenced to six months in prison and returned to Sainte-Pélagie. During the following months the teetotal youth was driven to drink by the rogues who surrounded him. The botanist and ardent republican François Raspail, who was imprisoned for refusing to accept the Cross of the Legion of Honour from Louis-Phillipe, wrote an account of Galois's first drinking bout:

He grasps the little glass like Socrates courageously taking the hemlock; he swallows it as one gulp, not without blinking and making a wry face. A second glass is not harder to empty than the first, and then the third. The beginner loses his equilibrium. Triumph! Homage to the Bacchus of the jail! You have intoxicated an ingenuous soul, who holds wine in horror.

A week later a sniper in a garret opposite the prison fired a shot into a cell wounding the man next to Galois. Galois was convinced that the bullet was intended for himself and that there was a government plot to assassinate him. The fear of political persecution terrorised him, and the isolation from his friends and family and rejection of his mathematical ideas plunged him into a state of depression. In a bout of drunken delirium he tried to stab himself to death, but Raspail and others managed to restrain and disarm him. Raspail recalls Galois's words immediately prior to the suicide attempt:

Do you know what I lack my friend? I confide it only to you: it is someone I can love and love only in spirit. I have lost my father and no one has ever replaced him, do you hear me . . . ?

In March 1832, a month before Galois's sentence was due to finish, a cholera epidemic broke out in Paris and the prisoners of Sainte-Pélagie were released. What happened to Galois over the next few weeks has been the subject of intense speculation, but what is certain is that the events of this period were largely the consequence of a romance with a mysterious woman by the name of Stéphanie-Félicie Poterine du Motel, the daughter of a respected Parisian physician. Although there are no clues as to how the affair started, the details of its tragic end are well documented.

Stéphanie was already engaged to a gentleman by the name of Pescheux d'Herbinville, who uncovered his fiancée's infidelity. D'Herbinville was furious and, being one the finest shots in France, he had no hesitation in immediately challenging Galois to a duel at dawn. Galois was well aware of his challenger's reputation. During the evening prior to the confrontation, which he believed would be his last opportunity to commit his thoughts to paper, he wrote letters to his friends explaining his circumstances:

I beg my patriots, my friends, not to reproach me for dying otherwise than for my country. I died the victim of an infamous coquette and her two

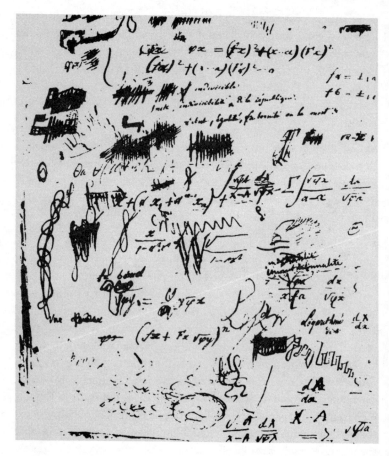

Figure 22(a). The night before the duel Galois attempted to write down all his mathematical ideas. Other comments, however, also appear in the notes. On this page, on the left below centre, are the words 'Une femme' with the second word scribbled out, presumably a reference to the woman at the centre of the duel.

dupes. It is in a miserable piece of slander that I end my life. Oh! Why die for something so little, so contemptible? I call on heaven to witness that only under compulsion and force have I yielded to a provocation which I have tried to avert by every means.

Despite his devotion to the republican cause and his romantic involvement, Galois had always maintained his passion for mathematics and one of his greatest fears was that his research, which had already been rejected by the Academy, would be lost forever. In a desperate attempt to gain recognition he worked through the night writing out the theorems which he believed fully explained the riddle of quintic equations. Figure 22 shows some of the last pages written by Galois. The pages were largely a transcription of the ideas he had already submitted to Cauchy and Fourier, but hidden within the complex algebra were occasional references to 'Stéphanie' or 'une femme' and exclamations of despair – 'I have not time, I have not time!' At the end of the night, when his calculations were complete, he wrote a covering letter to his friend Auguste Chevalier, requesting that, should he die, the papers be distributed to the greatest mathematicians in Europe:

My Dear Friend,

I have made some new discoveries in analysis. The first concern the theory of quintic equations, and others integral functions.

In the theory of equations I have researched the conditions for the solvability of equations by radicals; this has given me the occasion to deepen this theory and describe all the transformations possible on an equation even though it is not solvable by radicals. All this will be found here in three memoirs . . .

In my life I have often dared to advance propositions about which I was not sure. But all I have written down here has been clear in my head for over a year, and it would not be in my interest to leave myself open to the

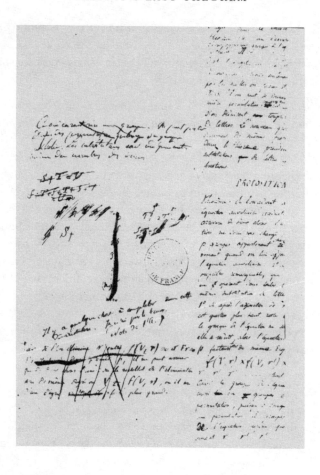

Figure 22(b) As Galois tried desperately to record everything before the fateful hour, it occurred to him that he might not complete the task. The words 'je n'ai pas le temps' (I have no time) are visible at the end of the two lines in the lower left part of the page.

suspicion that I announce theorems of which I do not have a complete proof.

Make a public request of Jacobi or Gauss to give their opinions, not as to the truth, but as to the importance of these theorems. After that, I hope some men will find it profitable to sort out this mess.

I embrace you with effusion,

E. Galois

The following morning, Wednesday 30 May 1832, in an isolated field Galois and d'Herbinville faced each other at twenty-five paces armed with pistols. D'Herbinville was accompanied by seconds; Galois stood alone. He had told nobody of his plight: a messenger he had sent to his brother Alfred would not deliver the news of the duel until it was over and the letters he had written the previous night would not reach his friends for several days.

The pistols were raised and fired. D'Herbinville still stood, Galois was hit in the stomach. He lay helpless on the ground. There was no surgeon to hand and the victor calmly walked away leaving his wounded opponent to die. Some hours later Alfred arrived on the scene and carried his brother to Cochin hospital. It was too late, peritonitis had set in, and the following day Galois died.

His funeral was almost as farcical as his father's. The police believed that it would be the focus of a political rally and arrested thirty comrades the previous night. Nonetheless two thousand republicans gathered for the service and inevitably scuffles broke out between Galois's colleagues and the government officials who had arrived to monitor events.

The mourners were angry because of a growing belief that d'Herbinville was not a cuckolded fiancé but rather a government agent, and that Stéphanie was not just a lover but a scheming

seductress. Events such as the shot which was fired at Galois while
he was in Sainte-Pélagie prison already hinted at a conspiracy to
assassinate the young trouble-maker, and therefore his friends concluded that he had been duped into a romance which was part of
a political plot contrived to kill him. Historians have argued about
whether the duel was the result of a tragic love affair or politically
motivated, but either way one of the world's greatest mathematicians was killed at the age of twenty, having studied mathematics
for only five years.

Before distributing Galois's papers his brother and Auguste
Chevalier rewrote them in order to clarify and expand the ex-
planations. Galois's habit of explaining his ideas hastily and inade-
quately was no doubt exacerbated by the fact that he had only a
single night to outline years of research. Although they dutifully
sent copies of the manuscript to Carl Gauss, Carl Jacobi and
others, there was no acknowledgment of Galois's work for over a
decade, until a copy reached Joseph Liouville in 1846. Liouville
recognised the spark of genius in the calculation and spent months
trying to interpret its meaning. Eventually he edited the papers and
published them in his prestigious *Journal de Mathématiques pures et
appliquées*. The response from other mathematicians was immediate and impressive because Galois had indeed formulated a complete understanding of how one could go about finding solutions to
quintic equations. First Galois had classified all quintics into two
types: those that were soluble and those that were not. Then, for
those that were soluble, he devised a recipe for finding the solutions
to the equations. Moreover, Galois examined equations of higher
order than the quintic, those containing x^6, x^7, and so on, and
could identify which of these were soluble. It was one of the
masterpieces of nineteenth-century mathematics created by one of
its most tragic heroes.

In his introduction to the paper Liouville reflected on why the young mathematician had been rejected by his seniors and how his own efforts had resurrected Galois:

An exaggerated desire for conciseness was the cause of this defect which one should strive above all else to avoid when treating the abstract and mysterious matters of pure Algebra. Clarity is, indeed, all the more necessary when one essays to lead the reader farther from the beaten path and into wilder territory. As Descartes said, 'When transcendental questions are under discussion be transcendentally clear.' Too often Galois neglected this precept; and we can understand how illustrious mathematicians may have judged it proper to try, by the harshness of their sage advice, to turn a beginner, full of genius but inexperienced, back on the right road. The author they censured was before them ardent, active; he could profit by their advice.

But now everything is changed. Galois is no more! Let us not indulge in useless criticisms; let us leave the defects there and look at the merits . . .

My zeal was well rewarded, and I experienced an intense pleasure at the moment when, having filled in some slight gaps, I saw the complete correctness of the method by which Galois proves, in particular, this beautiful theorem.

Toppling the First Domino

At the heart of Galois's calculations was a concept known as *group theory*, an idea which he had developed into a powerful tool capable of cracking previously insoluble problems. Mathematically, a group is a set of elements which can be combined together using some operation, such as addition or multiplication, and which satisfy certain conditions. An important defining property of a group

is that, when any two of its elements are combined using the operation, the result is another element in the group. The group is said
to be *closed* under that operation.

For example, the whole numbers form a group under the operation of 'addition'. Combining one whole number with another
under the operation of addition leads to a third whole number, e.g.

$$4 + 12 = 16.$$

All possible results under addition are within the whole numbers,
and so mathematicians state that 'the whole numbers are closed
under addition' or 'the whole numbers under addition form a
group'. On the other hand the whole numbers do *not* form a group
under the operation of 'division', because dividing one whole
number by another does not necessarily lead to another whole
number, e.g.

$$4 \div 12 = \tfrac{1}{3}.$$

The fraction $\tfrac{1}{3}$ is not a whole number and is outside the original
group. However, by considering a larger group which does include
fractions, the so-called rational numbers, closure can be re-established: 'the rational numbers are closed under division'. Having
said this, one stills needs to be careful because division by the
element zero results in infinity, which leads to various mathematical nightmares. For this reason it is more accurate to state that 'the
rational numbers (excluding zero) are closed under division'. In
many ways closure is similar to the concept of completeness described in earlier chapters.

The whole numbers and the fractions form infinitely large
groups, and one might assume that, the larger the group, the more
interesting the mathematics it will generate. However, Galois had
a 'less is more' philosophy, and showed that small carefully con-

structed groups could exhibit their own special richness. Instead of using the infinite groups, Galois began with a particular equation and constructed his group from the handful of solutions to that equation. It was groups formed from the solutions to quintic equations which allowed Galois to derive his results about these equations. A century and a half later Wiles would use Galois's work as the foundation for his proof of the Taniyama–Shimura conjecture.

To prove the Taniyama–Shimura conjecture, mathematicians had to show that every one of the infinite number of elliptic equations could be paired with a modular form. Originally they had attempted to show that the whole DNA for one elliptic equation (the E-series) could be matched with the whole DNA for one modular form (the M-series), and then they would move on to the next elliptic equation. Although this is a perfectly sensible approach, nobody had found a way to repeat this process over and over again for the infinite number of elliptic equations and modular forms.

Wiles tackled the problem in a radically different way. Instead of trying to match all elements of one E-series and M-series and then moving on to the next E-series and M-series, he tried to match one element of all E-series and M-series and then move on to the next element. In other words each E-series has an infinite list of elements, individual genes which make up the DNA, and Wiles wanted to show that the the first gene in every E-series could be matched with the first gene in every M-series. He would then go on to show that the second gene in every E-series could be matched with the second gene in every M-series, and so on.

In the traditional approach one had an infinite problem, which was that even if you could prove that all of one E-series matched all of one M-series, there were still infinitely many other E-series and M-series to be matched. Wiles's approach still involved tackling infinity because even if he could prove that the first gene of every E-series was identical to the first gene of every M-series there were still infinitely many other genes to be matched. However, Wiles's approach had one major advantage over the traditional approach.

In the old method, once you had proved that the whole of one E-series matched the whole of one M-series, you then had to ask, Which E-series and M-series do I try and match up next? The infinity of E-series and M-series have no natural order and so whichever one is tackled next is a largely arbitrary choice. Crucially, in Wiles's method, the genes in the E-series do have a natural order, and so having proved that all the first genes match ($E_1 = M_1$), the next step is obviously to prove that all the second genes match ($E_2 = M_2$), and so on.

This natural order is exactly what Wiles needed in order to develop an inductive proof. Initially Wiles would have to show that the first element of every E-series could be paired with the first element of every M-series. Then he would have to show that if the first elements could be paired then so could the second elements, and if the second elements could be paired then so could the third elements, and so on. He had to topple the first domino, and then he had to prove that any falling domino would also topple the next one.

The first step was achieved when Wiles realised the power of Galois's groups. A handful of solutions from every elliptic equation could be used to form a group. After months of analysis Wiles proved that the group led to one undeniable conclusion – the first element in every E-series could indeed be paired with the first one in an M-series. Thanks to Galois, Wiles had been able to topple the first domino. The next step of his inductive proof required him to find a way of showing that if any one element of the E-series matched the corresponding element in the M-series, then so must the next element match.

Getting this far had already taken two years, and there was no hint of how long it would take to find a way of extending the proof. Wiles was well aware of the task ahead: 'You might ask how could I devote an unlimited amount of time to a problem that might simply not be soluble. The answer is that I just loved working on this problem and I was obsessed. I enjoyed pitting my wits against it. Furthermore, I always knew that the mathematics I was thinking about, even if it wasn't strong enough to prove Taniyama–Shimura, and hence Fermat, would prove something. I wasn't going up a back alley, it was certainly good mathematics and that was true all along. There was certainly a possibility that I would never get to Fermat, but there was no question that I was simply wasting my time.'

'Fermat's Theorem Solved?'

Although it was only the first step towards proving the Taniyama–Shimura conjecture, Wiles's Galois strategy was a brilliant mathematical breakthrough, worthy of publication in its own right. As a result of his self-imposed seclusion he could not announce the result to the rest of the world, but similarly he had no idea who else might be making equally significant breakthroughs.

Wiles recalls his philosophical attitude towards any potential rivals: 'Well, obviously no one wants to spend years trying to solve something and then find that someone else just solves it a few weeks before you do. But curiously, because I was trying a problem that's considered impossible, I didn't really have much fear of competition. I simply didn't think I or anyone else had any real idea how to do it.'

On 8 March 1988 Wiles was shocked to read front-page head-lines announcing that Fermat's Last Theorem had been solved. The *Washington Post* and the *New York Times* claimed that thirty-eight-year-old Yoichi Miyaoka of the Tokyo Metropolitan University had discovered a solution to the world's hardest problem. At this stage Miyaoka had not yet published his proof, but only described its outline at a seminar at the Max Planck Institute for Mathematics in Bonn. Don Zagier who was in the audience summarised the community's optimism, 'Miyaoka's proof is very exciting and some people feel that there is a very good chance that it is going to work. It's still not definite, but it looks fine so far.'

In Bonn, Miyaoka had described how he had approached the problem from a completely new angle, namely differential geometry. For decades differential geometers had developed a rich understanding of mathematical shapes and in particular the properties of their surfaces. Then in the 1970s a team of Russians led by Professor S. Arakelov attempted to draw parallels between problems in differential geometry and problems in number theory. This was one strand of the Langlands programme, and the hope was that unanswered problems in number theory could be solved by examining the corresponding question in differential geometry which had already been answered. This was known as the *philosophy of parallelism*.

Differential geometrists who tried to tackle problems in number theory became known as 'arithmetic algebraic geometrists', and in 1983 they claimed their first significant victory, when Gerd Faltings at the Institute for Advanced Study at Princeton made a major contribution towards understanding Fermat's Last Theorem. Remember that Fermat claimed that there were no whole number solutions to the equation:

$$x^n + y^n = z^n \quad \text{for } n \text{ greater than 2.}$$

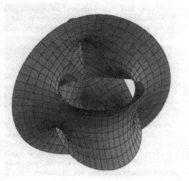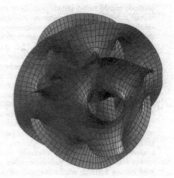

Figure 23. These surfaces were created using the computer program *Mathematica*. They are geometrical representations of the equation $x^n + y^n = 1$, where $n = 3$ for the first image and $n = 5$ for the second image. Here, x and y are regarded as complex variables.

Faltings believed he could make some progress towards proving the Last Theorem by studying the geometric shapes associated with different values of n. The shapes corresponding to each of the equations are all different, but they do have one thing in common – they are all punctured with holes. The shapes are four-dimensional, rather like modular forms, and a two-dimensional visualisation of two of them is shown in Figure 23. All the shapes are like multi-dimensional doughnuts, with several holes rather than just one. The larger the value of n in the equation, the more holes there are in the corresponding shape.

Faltings was able to prove that, because these shapes always have more than one hole, the associated Fermat equation could only have a finite number of whole number solutions. A finite number of solutions could be anything from zero, which was Fermat's own claim, to a million or a billion. So Faltings had not proved Fermat's Last Theorem, but he had at least been able to discount the possibility of an infinity of solutions.

Five years later Miyaoka claimed he could go one step further. While still in his early twenties he had created a conjecture concerning the so-called Miyaoka inequality. It became clear that proof of his own geometrical conjecture would demonstrate that the number of solutions for Fermat's equation was not only finite, but zero. Miyaoka's approach was analogous to Wiles's in that they were both trying to prove the Last Theorem by connecting it to a fundamental conjecture in a different field of mathematics. In Miyaoka's case it was differential geometry; for Wiles the proof was via elliptic equations and modular forms. Unfortunately for Wiles he was still struggling to prove the Taniyama–Shimura conjecture when Miyaoka announced a full proof relating to his own conjecture, and therefore a proof of Fermat's Last Theorem.

Two weeks after his announcement in Bonn, Miyaoka released the five pages of algebra which detailed his proof and then the scrutiny began. Number theorists and differential geometrists around the world examined the proof line by line, looking for the slightest gap in the logic or the merest hint of a false assumption. Within a few days several mathematicians highlighted what seemed to be a worrying contradiction within the proof. Part of Miyaoka's work led to a particular conclusion in number theory, which when translated back to differential geometry conflicted with a result which had already been proved years earlier. Although this did not necessarily invalidate Miyaoka's entire proof, it did clash with the philosophy of parallelism between number theory and differential geometry.

Another two weeks passed when Gerd Faltings, who had paved the way for Miyaoka, announced that he had pinpointed the exact reason for the apparent breakdown in parallelism – a gap in the logic. The Japanese mathematician was predominantly a geometrist and he had not been absolutely rigorous in translating his ideas into the less familiar territory of number theory. An army of number theorists attempted to help Miyaoka patch up the error but their efforts ended in failure. Two months after the initial announcement the consensus was that the original proof was destined to fail.

As with several other failed proofs in the past, Miyaoka had created new and interesting mathematics. Individual chunks of the proof stood on their own as ingenious applications of differential geometry to number theory, and in later years other mathematicians would build on them in order to prove other theorems, but never Fermat's Last Theorem.

The fuss over Fermat soon died down and the newspapers ran short updates explaining that the 300-year-old puzzle remained unsolved. No doubt inspired by all the media attention a new piece of graffiti found its way on to New York's Eighth Street subway station:

$$x^n + y^n = z^n: \quad \text{no solutions}$$

> I have discovered a truly remarkable proof of this,
> but I can't write it now because my train is coming.

The Dark Mansion

Unknown to the world Wiles breathed a sigh of relief. Fermat's Last Theorem remained unconquered and he could continue with his battle to prove it via the Taniyama–Shimura conjecture. 'Much of the time I would sit writing at my desk, but sometimes I could reduce the problem to something very specific – there's a clue, something that strikes me as strange, something just below the paper which I can't quite put my finger on. If there was one particular thing buzzing in my mind then I didn't need anything to write with or any desk to work at, so instead I would go for a walk down by the lake. When I'm walking I find I can concentrate my mind on one very particular aspect of a problem, focusing on it completely. I'd always have a pencil and paper ready, so if I had an idea I could sit down at a bench and start scribbling away.'

After three years of non-stop effort Wiles had made a series of breakthroughs. He had applied Galois groups to elliptic equations, he had broken the elliptic equations into an infinite number of pieces, and then he had proved that the first piece of every elliptic equation had to be modular. He had toppled the first domino and now he was exploring techniques which might lead to the collapse of all the others. In hindsight this seemed like the natural route to a proof, but getting this far had required enormous determination to overcome the periods of self-doubt. Wiles describes his experience of doing mathematics in terms of a journey through a dark unexplored mansion. 'One enters the first room of the mansion and it's dark. Completely dark. One stumbles around bumping into the furniture but gradually you learn where each piece of furniture is. Finally, after six months or so, you find the light switch, you turn it on, and suddenly it's all illuminated. You can see exactly where you were. Then you move into the next room and spend another six months in the dark. So each of these breakthroughs, while sometimes they're momentary, sometimes over a period of a day or two, they are the culmination of, and couldn't exist without, the many months of stumbling around in the dark that precede them.'

In 1990 Wiles found himself in what seemed to be the darkest room of all. He had been exploring it for almost two years. He still had no way of showing that if one piece of the elliptic equation was modular then so was the next piece. Having tried every tool and technique in the published literature, he had found that they were all inadequate. 'I really believed that I was on the right track, but that did not mean that I would necessarily reach my goal. It could be that the methods needed to solve this particular problem may simply be beyond present-day mathematics. Perhaps the methods I needed to complete the proof would not be invented for a hundred years. So even if I was on the right track, I could be living in the wrong century.'

Undaunted, Wiles persevered for another year. He began working on a technique called Iwasawa theory. Iwasawa theory was a method of analysing elliptic equations which he had learnt as a student in Cambridge under John Coates. Although the method as it stood was inadequate, he hoped that he could modify it and make it powerful enough to generate a domino effect.

Since making the initial breakthrough with Galois groups, Wiles had become increasingly frustrated. Whenever the pressure became too great, he would turn to his family. Since beginning work on Fermat's Last Theorem in 1986, he had become a father twice over. 'The only way I could relax was when I was with my children. Young children simply aren't interested in Fermat, they just want to hear a story and they're not going to let you do anything else.'

The Method of Kolyvagin and Flach

By the summer of 1991 Wiles felt he had lost the battle to adapt Iwasawa theory. He had to prove that every domino, if it itself had been toppled, would topple the next domino – that if one element in the elliptic equation E-series matched one element in the modular form M-series, then so would the next one. He also had to be sure that this would be the case for every elliptic equation and every modular form. Iwasawa theory could not give him the guarantee he required. He completed another exhaustive search of the literature and was still unable to find an alternative technique which would give him the breakthrough he needed. Having been a virtual recluse in Princeton for the last five years, he decided it was time to get back into circulation in order to find out the latest mathematical gossip. Perhaps somebody somewhere was working on an innovative new technique, and as yet, for whatever reason, had not published it. He headed north to Boston to attend a major conference on elliptic equations, where he would be sure of meeting the major players in the subject.

Wiles was welcomed by colleagues from around the world, who were delighted to see him after such a long absence from the conference circuit. They were still unaware of what he had been working on and Wiles was careful not to give away any clues. They did not suspect his ulterior motive when he asked them the latest news concerning elliptic equations. Initially the responses were of no relevance to Wiles's plight but an encounter with his former supervisor John Coates was more fruitful: 'Coates mentioned to me that a student of his named Matheus Flach was writing a beautiful paper in which he was analysing elliptic equations. He was building on a recent method devised by Kolyvagin and it looked as though his method was tailor-made for my problem. It seemed to be exactly what I needed, although I knew I would still have to further develop this so-called Kolyvagin–Flach method. I put aside completely the old approach I'd been trying and I devoted myself night and day to extending Kolyvagin–Flach.'

In theory this new method could extend Wiles's argument from the first piece of the elliptic equation to all pieces of the elliptic equation, and potentially it could work for evey elliptic equation. Professor Kolyvagin had devised an immensely powerful mathematical method, and Matheus Flach had refined it to make it even more potent. Neither of them realised that Wiles intended to incorporate their work into the world's most important proof.

Wiles returned to Princeton, spent several months familiarising himself with his newly discovered technique, and then began the mammoth task of adapting it and implementing it. Soon for a particular elliptic equation he could make the inductive proof work – he could topple all the dominoes. Unfortunately the Kolyvagin–Flach method that worked for one particular elliptic equation did not necessarily work for another elliptic equation. He eventually realised that all the elliptic equations could be classified into various families. Once modified to work on one elliptic equation, the Kolyvagin–Flach method would work for all the other elliptic equations in that family. The challenge was to adapt the Kolyvagin–Flach method to work for each family. Although some families were harder to conquer than others, Wiles was confident that he could work his way through them one by one.

After six years of intense effort Wiles believed that the end was in sight. Week after week he was making progress, proving that newer and bigger families of elliptic curves must be modular. It seemed to be just a question of time before he would mop up the outstanding elliptic equations. During this final stage of the proof, Wiles began to appreciate that his whole proof relied on exploiting a technique which he had only discovered a few months earlier. He began to question whether he was using the Kolyvagin–Flach method in a fully rigorous manner.

'During that year I worked extremely hard trying to make the Kolyvagin–Flach method work, but it involved a lot of sophisti-

Nick Katz

cated machinery that I wasn't really familiar with. There was a lot of hard algebra which required me to learn a lot of new mathematics. Then around early January of 1993 I decided that I needed to confide in someone who was an expert in the kind of geometric techniques I was invoking for this. I wanted to choose very carefully who I told because they would have to keep it confidential. I chose to tell Nick Katz.'

Professor Nick Katz also worked in Princeton University's Mathematics Department and had known Wiles for several years. Despite their closeness Katz was oblivious to what was going on literally just along the corridor. He recalls every detail of the moment Wiles revealed his secret: 'One day Andrew came up to me at tea and asked me if I could come up to his office – there was something he wanted to talk to me about. I had no idea of what this could be. I went up to his office and he closed the door. He said he thought that he would be able to prove the Taniyama–Shimura conjecture. I was just amazed, flabbergasted – this was fantastic.

'He explained that there was a big part of the proof that relied on his extension of the work of Flach and Kolyvagin but it was pretty technical. He really felt shaky on this highly technical part of the proof and he wanted to go through it with somebody because he wanted to be sure it was correct. He thought I was the right person to help him check it, but I think there was another reason why he asked me in particular. He was sure that I would keep my mouth shut and not tell other people about the proof.'

After six years in isolation Wiles had let go of his secret. Now it was Katz's job to get to grips with a mountain of spectacular calculations based on the Kolyvagin–Flach method. Virtually everything Wiles had done was revolutionary and Katz gave a great deal of thought as to the best way to examine it thoroughly: 'What Andrew had to explain was so big and long that it wouldn't have worked to try and just explain it in his office in informal conversations. For something this big we really needed to have the formal structure of weekly scheduled lectures, otherwise the thing would just degenerate. So, that's why we decided to set up a lecture course.'

They decided that the best strategy would be to announce a series of lectures open to the department's graduate students. Wiles would give the course and Katz would be in the audience. The course would effectively cover the part of the proof that needed checking but the graduate students would have no idea of this. The beauty of disguising the checking of the proof in this way was that it would force Wiles to explain everything step by step, and yet it would not arouse any suspicion within the department. As far as everyone else was concerned this was just another graduate course.

'So Andrew announced this lecture course called "Calculations on Elliptic Curves",' recalls Katz with a sly smile, 'which is a completely innocuous title – it could mean anything. He didn't mention Fermat, he didn't mention Taniyama–Shimura, he just started by diving right into doing technical calculations. There was no way in the world that anyone could have guessed what it was really about. It was done in such a way that unless you knew what this was for, then the calculations would just seem incredibly technical and tedious. And when you don't know what the mathematics is for, it's impossible to follow it. It's pretty hard to follow it even when you do know what it's for. Anyway, one by one the graduate students just drifted away and after a few weeks I was the only person left in the audience.'

Katz sat in the lecture theatre and listened carefully to every step of Wiles's calculation. By the end of it his assessment was that the Kolyvagin–Flach method seemed to be working perfectly. Nobody else in the department realised what had been going on. Nobody

suspected that Wiles was on the verge of claiming the most important prize in mathematics. Their plan had been a success.

Once the lecture series was over Wiles devoted all his efforts to
completing the proof. He had successfully applied the Kolyvagin–
Flach method to family after family of elliptic equations, and by
this stage only one family refused to submit to the technique. Wiles
describes how he attempted to complete the last element of the
proof: 'One morning in late May, Nada was out with the children
and I was sitting at my desk thinking about the remaining family of
elliptic equations. I was casually looking at a paper of Barry
Mazur's and there was one sentence there that just caught my
attention. It mentioned a nineteenth-century construction, and I
suddenly realised that I should be able to use that to make the
Kolyvagin–Flach method work on the final family of elliptic equations. I went on into the afternoon and I forgot to go down for
lunch, and by about three or four o'clock I was really convinced
that this would solve the last remaining problem. It got to about
tea-time and I went downstairs and Nada was very surprised that
I'd arrived so late. Then I told her – I'd solved Fermat's Last
Theorem.'

The Lecture of the Century

After seven years of single-minded effort Wiles had completed a
proof of the Taniyama–Shimura conjecture. As a consequence,
and after thirty years of dreaming about it, he had also proved
Fermat's Last Theorem. It was now time to tell the rest of the
world.

'So by May 1993, I was convinced that I had the whole of
Fermat's Last Theorem in my hands,' recalls Wiles. 'I still wanted

to check the proof some more but there was a conference which was coming up at the end of June in Cambridge, and I thought that would be a wonderful place to announce the proof – it's my old home town, and I'd been a graduate student there.'

The conference was being held at the Isaac Newton Institute. This time the institute had planned a workshop on number theory with the obscure title 'L-functions and Arithmetic'. One of the organisers was Wiles's Ph.D. supervisor John Coates: 'We brought people from all around the world who were working on this general circle of problems and, of course, Andrew was one of the people that we invited. We'd planned one week of concentrated lectures and originally, because there was a lot of demand for lecture slots, I only gave Andrew two lecture slots. But then I gathered he needed a third slot, and so in fact I arranged to give up my own slot for his third lecture. I knew that he had some big result to announce but I had no idea what.'

When Wiles arrived in Cambridge he had two and a half weeks before his lectures began and wanted to make the most of the opportunity: 'I decided I would check the proof with one or two experts, in particular the Kolyvagin–Flach part. The first person I gave it to was Barry Mazur. I think I said to him, "I have a manuscript here with a proof to a certain theorem." He looked very baffled for a while, and then I said, "Well, have a look at it." I think it then took him some time to register. He appeared stunned. Anyway I told him that I was hoping to speak about it at the conference, and that I'd really like him to try and check it.'

One by one the most eminent figures in number theory began to arrive at the Newton Institute, including Ken Ribet whose calculation in 1986 had inspired Wiles's seven-year ordeal. 'I arrived at this conference on L-functions and elliptic curves and it didn't seem to be anything out of the ordinary until people started telling me that they had been hearing weird rumours about Andrew Wiles's proposed series of lectures. The rumour was that he had proved Fermat's Last Theorem, and I just thought this was completely nuts. I thought it couldn't possibly be true. There are lots of cases when rumours start circulating in mathematics, especially through electronic mail, and experience shows that you shouldn't put too much stock in them. But the rumours were very persistent and Andrew was refusing to answer questions about it and he was behaving very very queerly. John Coates said to him, "Andrew, what have you proved? Shall we call the press?" Andrew just kind of shook his head and sort of kept his lips sealed. He was really going for high drama.

'Then one afternoon Andrew came up to me and started asking me about what I'd done in 1986 and some of the history of Frey's ideas. I thought to myself, this is incredible, he must have proved the Taniyama–Shimura conjecture and Fermat's Last Theorem, otherwise he wouldn't be asking me this. I didn't ask him directly if this was true, because I saw that he was behaving very coyly and I knew I wouldn't get a straight answer. So I just kind of said, "Well Andrew, if you have occasion to speak about this work, here's what happened." I sort of looked at him as though I knew something, but I didn't really know what was going on. I was still just guessing.'

Wiles's reaction to the rumours and the mounting pressure was simple: 'People would ask me, leading up to my lectures, what exactly I was going to say. So I said, well, come to my lectures and see.'

Back in 1920 David Hilbert, then aged fifty-eight, gave a public lecture in Göttingen on the subject of Fermat's Last Theorem. When asked if the problem would ever be solved, he replied that he would not live to see it, but perhaps younger members of the audience might witness the solution. Hilbert's estimate for the date of the solution was proving to be fairly accurate. Wiles's lecture was also well timed in relation to the Wolfskehl Prize. In his will Paul Wolfskehl had set a deadline of 13 September 2007.

The title of Wiles's lecture series was 'Modular Forms, Elliptic Curves and Galois Representations'. Once again, as with the graduate lectures he had given earlier in the year for the benefit of Nick Katz, the title of the lectures was so vague that it gave no hint of his ultimate aim. Wiles's first lecture was apparently mundane, laying the foundations for his attack on the Taniyama–Shimura conjecture in the second and third. The majority of his audience were completely unaware of the gossip, did not appreciate the point of the lectures, and paid little attention to the details. Those in the know were looking for the slightest clue which might give credence to the rumours.

Immediately after the lecture ended the rumour mill started again with renewed vigour, and electronic mail flew around the world. Professor Karl Rubin, a former student of Wiles, reported back to his colleagues in America:

```
Date:          Mon, 21 Jun 1993 13:33:06
Subject:       Wiles
```

```
Hi.            Andrew gave his first talk today. He
did not announce a proof of Taniyama—Shimura, but
he is moving in that direction and he has two
more lectures. He is still being very secretive
about the final result.

My best guess is that he is going to prove that
if E is an elliptic curve over Q and the Galois
representation on the points of order 3 on E
satisfies certain hypotheses, then E is modular.
```

From what he has said it seems he will not prove
the full conjecture. What I don't know is whether
this will apply to Frey's curve, and therefore
say something about Fermat. I'll keep you posted.

Karl Rubin
Ohio State University

By the following day more people had heard the gossip, and so the
audience for the second lecture was significantly larger. Wiles
teased them with an intermediate calculation which showed that
he was clearly trying to tackle the Taniyama–Shimura conjecture,
but the audience was still left wondering if he had done enough to
prove it and, as a consequence, conquer Fermat's Last Theorem.
A new batch of e-mails bounced off the satellites.

Date: Tue, 22 Jun 1993 13:10:39
Subject: Wiles

No more real news in today's lecture. Andrew
stated a general theorem about lifting Galois
representations along the lines I suggested
yesterday. It does not seem to apply to all
elliptic curves but the punchline will come
tomorrow.

I don't really know why he is doing it this way.
It's clear that he knows what he is going to say
tomorrow. This is a truly massive piece of work
that he has been working on for years, and he
seems confident of it. I'll let you know what
happens tomorrow.

Karl Rubin
Ohio State University

'On 23 June Andrew began his third and final lecture,' recalls John Coates. 'What was remarkable was that practically everyone who contributed to the ideas behind the proof was there in the room, Mazur, Ribet, Kolyvagin, and many, many others.'

By this point the rumour was so persistent that everyone from the Cambridge mathematics community turned up for the final lecture. The lucky ones were crammed into the auditorium, while the others had to wait in the corridor, where they stood on tip-toe and peered through the window. Ken Ribet had made sure that he would not miss out on the most important mathematical announcement of the century: 'I came relatively early and I sat in the front row along with Barry Mazur. I had my camera with me just to record the event. There was a very charged atmosphere and people were very excited. We certainly had the sense that we were participating in a historical moment. People had grins on their faces before and during the lecture. The tension had built up over the course of several days. Then there was this marvellous moment when we were coming close to a proof of Fermat's Last Theorem.'

Barry Mazur had already been given a copy of the proof by Wiles, but even he was astonished by the performance. 'I've never seen such a glorious lecture, full of such wonderful ideas, with such dramatic tension, and what a build-up. There was only one possible punch line.'

After seven years of intense effort Wiles was about to announce his proof to the world. Curiously Wiles cannot remember the final moments of the lecture in great detail, but does recall the atmosphere: 'Although the press had already got wind of the lecture, fortunately they were not at the lecture. But there were plenty of people in the audience who were taking photographs towards the end and the Director of the Institute certainly had come well prepared with a bottle of champagne. There was a typical dignified silence while I read out the proof and then I just wrote up the statement of Fermat's Last Theorem. I said, "I think I'll stop here", and then there was sustained applause.'

The Aftermath

Strangely, Wiles was ambivalent about the lecture: 'It was obviously a great occasion, but I had mixed feelings. This had been part of me for seven years: it had been my whole working life. I got so wrapped up in the problem that I really felt I had it all to myself, but now I was letting go. There was a feeling that I was giving up a part of me.'

Wiles's colleague Ken Ribet had no such qualms: 'It was a completely remarkable event. I mean, you go to a conference and there are some routine lectures, there are some good lectures and there are some very special lectures, but it's only once in a lifetime that you get a lecture where someone claims to solve a problem that has endured for 350 years. People were looking at each other and saying, "My God, you know we've just witnessed an historical event." Then people asked a few questions about technicalities of the proof and possible applications to other equations, and then there was more silence and all of a sudden a second round of applause. The next talk was given by one Ken Ribet, yours truly. I gave the lecture, people took notes, people applauded, and no one present, including me, has any idea what I said in that lecture.'

While mathematicians were spreading the good news via e-mail, the rest of the world had to wait for the evening news, or the following day's newspapers. TV crews and science reporters descended upon the Newton Institute, all demanding interviews with the 'greatest mathematician of the century'. The *Guardian* exclaimed, 'The Number's Up for Maths' Last Riddle', and the front page of *Le Monde* read, 'Le théorèm de Fermat enfin résolu'. Journalists everywhere asked mathematicians for their expert opinion on Wiles's work, and professors, still recovering from the shock, were expected to briefly explain the most complicated mathematical proof ever, or provide a soundbite which would clarify the Taniyama–Shimura conjecture.

The first time Professor Shimura heard about the proof of his own conjecture was when he read the front page of the *New York Times* – 'At Last, Shout of "Eureka!" In Age-Old Math Mystery'. Thirty-five years after his friend Yutaka Taniyama had committed suicide, the conjecture which they had created together had now been vindicated. For many professional mathematicians the proof of the Taniyama–Shimura conjecture was a far more important achievement than the solution of Fermat's Last Theorem, because it had immense consequences for many other mathematical theorems. The journalists covering the story tended to concentrate on Fermat and mentioned Taniyama–Shimura only in passing, if at all.

Shimura, a modest and gentle man, was not unduly bothered by the lack of attention given to his role in the proof of Fermat's Last Theorem, but he was concerned that he and Taniyama had been relegated from being nouns to adjectives. 'It is very curious that people write about the Taniyama–Shimura conjecture, but nobody writes about Taniyama and Shimura.'

This was the first time that mathematics had hit the headlines since Yoichi Miyaoka announced his so-called proof in 1988: the only difference this time was that there was twice as much coverage and nobody expressed any doubt over the calculation. Overnight Wiles became the most famous, in fact the only famous, mathematician in the world, and *People* magazine even listed him

Following Wiles's lecture, newspapers around the world reported his proof of Fermat's Last Theorem.

among 'The 25 most intriguing people of the year', along with Princess Diana and Oprah Winfrey. The ultimate accolade came from an international clothing chain who asked the mild-mannered genius to endorse their new range of menswear.

While the media circus continued and while mathematicians made the most of being in the spotlight, the serious work of checking the proof was under way. As with all scientific disciplines each new piece of work has to be throughly examined, before it could be accepted as accurate and correct. Wiles's proof had to be submitted to the ordeal of trial by referee. Although Wiles's lectures at the Isaac Newton Institute had provided the world with an outline of his calculation, this did not qualify as official peer review. Academic protocol demands that any mathematician submits a complete manuscript to a respected journal, the editor of which then sends it to a team of referees whose job it is to examine the proof line by line. Wiles had to spend the summer anxiously waiting for the referees' report, hoping that eventually he would get their blessing.

Andrew Wiles and Ken Ribet immediately following the historic lecture at the Isaac Newton Institute.

7

A Slight Problem

A problem worthy of attack
Proves its worth by fighting back.
                                        Piet Hein

As soon as the Cambridge lecture was over, the Wolfskehl committee was informed of Wiles's proof. They could not award the prize immediately because the rules of the contest clearly demand verification by other mathematicians and official publication of the proof:

The *Königliche Gesellschaft der Wissenschaften* in Göttingen . . . will only take into consideration those mathematical memoirs which have appeared in the form of a monograph in the periodicals, or which are for sale in the bookshops . . . The award of the Prize by the Society will take place not earlier than two years after the publication of the memoir to be crowned. The interval of time is intended to allow German and foreign mathematicians to voice their opinion about the validity of the solution published.

Wiles submitted his manuscript to the journal *Inventiones Mathematicae*, whereupon its editor Barry Mazur began the process of selecting the referees. Wiles's paper involved such a variety of mathematical techniques, both ancient and modern, that Mazur made the exceptional decision to appoint not just two or three referees, as is usual, but six. Each year thirty thousand papers are published in journals around the world, but the sheer size and importance of Wiles's manuscript meant that it would undergo a unique level of scrutiny. To simplify matters the 200-page proof was divided into six sections and each of the referees took responsibility for one of these chapters.

Chapter 3 was the responsibility of Nick Katz, who had already examined that part of Wiles's proof earlier in the year: 'I happened to be in Paris for the summer to work at the Institut des Hautes Etudes Scientifique, and I took with me the complete 200-page proof – my particular chapter was seventy pages long. When I got there I decided I wanted to have serious technical help, and so I insisted that Luc Illusie, who was also in Paris, become a joint referee on this chapter. We would meet a few times a week throughout that summer, basically lecturing to each other to try and understand this chapter. Literally we did nothing but look through this manuscript line by line to try and make sure that there were no mistakes. Sometimes we got confused by things and so every day, sometimes twice a day, I would e-mail Andrew with a question – I don't understand what you say on this page or it seems to be wrong on this line. Typically I would get a response that day or the next day which clarified the matter and then we'd go on to the next problem.'

The proof was a gigantic argument, intricately constructed from hundreds of mathematical calculations glued together by thousands of logical links. If just one of the calculations was flawed or if one of the links became unstuck then the entire proof was potentially worthless. Wiles, who was now back in Princeton, anxiously waited for the referees to complete their task. 'I don't like to celebrate full out until I have the paper completely off my hands. In the meantime I had my work cut out dealing with the questions I was getting via e-mail from the referees. I was still pretty confident that none of these questions would cause me much trouble.' He had already checked and double-checked the proof before releasing it to the referees, so he was expecting little more than the mathematical equivalent of grammatical or typographic errors, trivial mistakes which he could fix immediately.

'These questions continued relatively uneventfully through till August,' recalls Katz, 'until I got to what seemed like just one more little problem. Sometime around 23 August I e-mail Andrew, but it's a little bit complicated so he sends me back a fax. But the fax doesn't seem to answer the question so I e-mail him again and I get another fax which I'm still not satisfied with.'

Wiles had assumed that this error was as shallow as all the others, but Katz's persistence forced him to take it seriously: 'I couldn't immediately resolve this one very innocent looking question. For a little while it seemed to be of the same order as the other problems, but then sometime in September I began to realise that this wasn't just a minor difficulty but a fundamental flaw. It was an error in a crucial part of the argument involving the Kolyvagin–Flach method, but it was something so subtle that I'd missed it completely until that point. The error is so abstract that it can't really be described in simple terms. Even explaining it to a mathematician would require the mathematician to spend two or three months studying that part of the manuscript in great detail.'

In essence the problem was that there was no guarantee that the Kolyvagin–Flach method worked as Wiles had intended. It was supposed to extend the proof from the first element of all elliptic equations and modular forms to cover all the elements, providing the toppling mechanism from one domino to the next. Originally the Kolyvagin–Flach method only worked under particularly constrained circumstances, but Wiles believed he had adapted and strengthened it sufficiently to work for all his needs. According to Katz this was not necessarily the case, and the effects were dramatic and devastating.

The error did not necessarily mean that Wiles's work was beyond salvation, but it did mean that he would have to strengthen his proof. The absolutism of mathematics demanded that Wiles demonstrate beyond all doubt that his method worked for every element of every E-series and M-series.

The Carpet Fitter

When Katz realised the significance of the error which he had spotted, he began to ask himself how he had missed it in the spring when Wiles had lectured to him with the sole purpose of identifying any mistakes. 'I think the answer is that there's a real tension when you're listening to a lecture between understanding everything and letting the lecturer get on with it. If you interrupt every second – I don't understand this or I don't understand that – then the guy never gets to explain anything and you don't get anywhere. On the other hand if you never interrupt you just sort of get lost and you're nodding your head politely, but you're not really checking anything. There's this real tension between asking too many questions and asking too few, and obviously by the end of those lectures, which is where this problem slipped through, I had erred on the side of too few questions.'

Only a few weeks earlier, newspapers around the globe had dubbed Wiles the most brilliant mathematician in the world, and after 350 years of frustration number theorists believed that they had at last got the better of Pierre de Fermat. Now Wiles was faced with the humiliation of having to admit that he had made a mistake. Before confessing to the error he decided to try and make a concerted effort to fill in the gap. 'I couldn't give up. I was obsessed by this problem and I still believed that the Kolyvagin–Flach method just needed a little tinkering. I just needed to modify it in some small way and then it would work just fine. I decided to go straight back into my old mode and completely shut myself off from the outside world. I had to focus again but this time under much more difficult circumstances. For a long time I would think that the fix was just round the corner, that I was just missing something simple and it would all just fit into place the next day. Of course it could have happened that way, but as time went by it seemed that the problem just became more intransigent.'

The hope was that he could fix the mistake before the mathematical community was aware that a mistake even existed. Wiles's wife, who had already witnessed the seven years of effort that had gone into the original proof, now had to watch her husband's agonising struggle with an error that could destroy everything. Wiles remembers her optimism: 'In September Nada said to me that the only thing she wanted for her birthday was a correct proof. Her birthday is on 6 October. I had only two weeks to deliver the proof, and I failed.'

For Nick Katz, too, this was a tense period: 'By October the only people who knew about the error were myself, Illusie, the other referees of other chapters and Andrew – in principle that was all. My attitude was that as a referee I was supposed to act in confidentiality. I certainly didn't feel that it was my business to discuss this matter with anyone except Andrew, so I just didn't say a word about it. I think externally he appeared normal but at this point he was keeping a secret from the world, and I think he must have been pretty uncomfortable about it. Andrew's attitude was that with just another day he would solve it, but as the fall went on, and no manuscript was available, rumours started circulating that there was a problem.'

In particular, Ken Ribet, another of the referees, began to feel the pressure of keeping the secret: 'For some completely accidental reason I became known as the "Fermat Information Service". There was an initial article in the *New York Times*, where Andrew asked me to speak to the reporter in his place, and the article said, 'Ribet who is acting as a spokesperson for Andrew Wiles . . .', or something to that effect. After that I became a magnet for all kinds of interest in Fermat's Last Theorem, both from inside and outside the mathematics community. People were calling from the press, from all around the world in fact, and also I gave a very large number of lectures over a period of two or three months. In these lectures I stressed what a magnificent achievement this was and I outlined the proof and I talked about the parts that I knew best, but after a while people started getting impatient and began asking awkward questions.

'You see Wiles had made this very public announcement, but no one outside of the very small group of referees had seen a copy of the manuscript. So mathematicians were waiting for this manuscript that Andrew had promised a few weeks after the initial announcement in June. People said, "Okay, well this theorem has been announced – we'd like to see what's going on. What's he doing? Why don't we hear anything?" People were a little upset that they were being held in ignorance and they simply wanted to know what was going on. Then things got even worse because slowly this cloud gathered over the proof and people kept telling me about these rumours, which claimed there was a gap in chapter 3. They'd ask me what I knew about it, and I just didn't know what to say.'

With Wiles and the referees denying any knowledge of a gap, or at the very least refusing to comment, speculation began to run wild. In desperation mathematicians began sending e-mails to each other in the hope of getting to the bottom of the mystery.

```
Subject:     Gap in Wiles proof?
Date:        18 Nov 1993 21:04:49 GMT

There are many rumours buzzing around about one
or more gaps in Wiles' proof. Does gap mean
crack, fissure, crevasse, chasm, or abyss? Does
anyone have reliable information?

Joseph Lipman
Purdue University
```

In every tea-room of every mathematics department the gossip surrounding Wiles's proof escalated every day. In response to the rumours and the speculative e-mails some mathematicians tried to return a sense of calm to the community.

```
Subject:     Reply: Gap in Wiles proof?
Date:        19 Nov 1993 15:42:20 GMT

I don't have any first hand information, and I
don't feel at liberty to discuss second hand
information. I think the best advice for everyone
is to keep calm and let the very competent
referees who are carefully examining Wiles' paper
do their work. They will report their findings
when they have something definite to say. Anyone
who has written a paper or refereed a paper will
be familiar with the fact that often questions
will arise in the process of checking proofs. It
```

would be amazing were this not to happen for such
an important result with a long difficult proof.

Leonard Evens
North Western University

Despite the calls for calm, the e-mails continued unabated. As well
as discussing the putative error, mathematicians were now arguing
over the ethics of pre-empting the referees' announcement.

Subject: More Fermat Gossip

Date: 24 Nov 93 12:00:34 GMT

I guess it's clear that I disagree with those who
say we should not gossip about whether Wiles'
proof of Fermat's Last Theorem is flawed or not. I
am all in favor of this kind of gossip as long as
it is not taken too seriously. I don't regard it
as malicious. In particular because, whether or
not Wiles' proof is flawed, I feel sure that he
has done some world-class mathematics.

So, here is what I got today, nth-hand...
Bob Silverman

Subject: Re: Fermat hole
Date: Mon, 22 Nov 93 20:16 GMT

Coates said in a lecture at the Newton Institute
here last week that in his opinion there is a gap
in the "geometric Euler systems" part of the proof
which "might take a week, or might take two years"
to fill. I have spoken to him several times, but am

still not sure on what basis he makes the claim:
he does not have a copy of the manuscript.

As far as I know the only copy in Cambridge is
with Richard Taylor as one of the referees of the
paper for Inventiones, and he has consistently
declined to comment until all the referees reach
a common conclusion. So the situation is
confused. Myself I don't see how Coates' view can
be taken as authoritative at this stage: I plan
to wait for word from Richard Taylor.

Richard Pinch

While the furore over his elusive proof was increasing, Wiles did his best to ignore the controversy and speculation. 'I really shut myself off because I didn't want to know what people were saying about me. I just went into seclusion but periodically my colleague Peter Sarnak would say to me, "You know that there's a storm out there." I'd listen, but, for myself, I really just wanted to cut myself off completely, just to focus completely on the problem.'

Peter Sarnak had joined the Princeton Mathematics Department at the same time as Wiles, and over the years they had become close friends. During this intense period of turmoil Sarnak was one of the few people in whom Wiles would confide. 'Well, I never knew the exact details, but it was clear that he was trying to overcome this one serious issue. But every time he would fix this one part of the calculation, it would cause some other difficulty in another part of the proof. It was like he was trying to put a carpet in a room where the carpet might be bigger than the room. So Andrew could fit the carpet in any one corner, only to find that it would pop up in another corner. Whether you could or could not fit the carpet in the room was not something he was able to decide. Mind you, even with the error, Andrew had made a giant step. Before him there was no one who had any approach to the Taniyama–Shimura conjecture, but now everybody got really excited because he showed us so many new ideas. They were fundamental, new things that nobody had considered before. So even if it couldn't be fixed this was a very major advance – but of course Fermat would still be unsolved.'

Eventually Wiles realised that he could not maintain his silence forever. The solution to the mistake was not just round the corner, and it was time to put an end to the speculation. After an autumn of dismal failure he sent the following e-mail to the mathematical bulletin board:

```
Subject:      Fermat Status
Date:         4 Dec 93 01:36:50 GMT

In view of the speculation on the status of my
work on the Taniyama-Shimura conjecture and
Fermat's Last Theorem I will give a brief account
of the situation. During the review process a
number of problems emerged, most of which have
been resolved, but one in particular I have not
settled. The key reduction of (most cases of) the
Taniyama-Shimura conjecture to the calculation of
the Selmer group is correct. However the final
calculation of a precise upper bound for the
Selmer group in the semistable case (of the
symmetric square representation associated to a
modular form) is not yet complete as it stands. I
believe that I will be able to finish this in the
```

near future using the ideas explained in my
Cambridge lectures.

The fact that a lot of work remains to be done on
the manuscript makes it still unsuitable for
release as a preprint. In my course in Princeton
beginning in February I will give a full account
of this work.

Andrew Wiles

Few were convinced by Wiles's optimism. Almost six months had passed without the error being corrected, and there was no reason to think anything would change in the next six months. In any case, if he really could 'finish this in the near future', then why bother issuing the e-mail? Why not just maintain the silence for a few more weeks and then release the finished manuscript? The February lecture course which he mentioned in his e-mail failed to give any of the promised detail, and the mathematical community suspected that Wiles was just trying to buy himself extra time.

The newspapers leapt on the story once again and mathematicians were reminded of Miyaoka's failed proof in 1988. History was repeating itself. Number theorists were now waiting for the next e-mail which would explain why the proof was irretrievably flawed. A handful of mathematicians had expressed doubts over the proof back in the summer, and now their pessimism seemed to have been justified. One story claims that Professor Alan Baker at the University of Cambridge offered to bet one hundred bottles of wine against a single bottle that the proof would be shown to be invalid within a year. Baker denies the anecdote, but proudly admits to having expressed a 'healthy scepticism'.

Less than six months after his lecture at the Newton Institute

Wiles's proof was in tatters. The pleasure, passion and hope that carried him through the years of secret calculations were replaced with embarrassment and despair. He recalls how his childhood dream had become a nightmare: 'The first seven years that I worked on this problem I enjoyed the private combat. No matter how hard it had been, no matter how insurmountable things seemed, I was engaged in my favourite problem. It was my childhood passion, I just couldn't put it down, I didn't want to leave it for a moment. Then I'd spoken about it publicly, and in speaking about it there was actually a certain sense of loss. It was a very mixed emotion. It was wonderful to see other people reacting to the proof, to see how the arguments could completely change the whole direction of mathematics, but at the same time I'd lost that personal quest. It was now open to the world and I no longer had this private dream which I was fulfilling. And then, after there was a problem with it, there were dozens, hundreds, thousands of people who wanted to distract me. Doing maths in that kind of rather overexposed way is certainly not my style and I didn't at all enjoy this very public way of doing it.'

Number theorists around the world empathised with Wiles's position. Ken Ribet had himself been through the same nightmare eight years earlier when he tried to prove the link between the Taniyama–Shimura conjecture and Fermat's Last Theorem. 'I was giving a lecture about the proof at the Mathematical Sciences Research Institute in Berkeley and someone from the audience said, "Well, wait a minute, how do you know that such and such is true?" I responded immediately giving my reason and they said, "Well that doesn't apply in this situation." I had an immediate terror. I kind of broke out into a sweat and I was very upset about it. Then I realised that there was only one possibility for justifying this, which was to go back to the fundamental work on the subject and see exactly how it was done in a similar situation. I looked in the relevant paper and I saw that the method did indeed apply in my case, and within a day or two I had the thing all fixed up. In my next lecture I was able to give the justification. But you always live with this fear that if you announce something important, a fundamental mistake can be discovered.

'When you find an error in a manuscript it can go two ways. Sometimes there's an immediate confidence and the proof can be resurrected with little difficulty. And sometimes there's the opposite. It's very disquieting, there's a sinking feeling when you realise that you've made a fundamental error and there's no way to repair it. It's possible that when a hole develops the theorem really just falls apart completely, because the more you try to patch it the more trouble you get into. But in Wiles's case each chapter of the proof was a significant article in its own right. The manuscript was seven years' work, it was basically several important papers pieced together and each one of the papers has a great deal of interest. The error occurred in one of the papers, in chapter 3, but even if you take out chapter 3 what remained was absolutely wonderful.'

But without chapter 3 there was no proof of the Taniyama–Shimura conjecture and therefore no proof of Fermat's Last Theorem. There was a sense of frustration in the mathematical community that the proof behind two great problems was in jeopardy. Moreover, after six months of waiting still nobody, beyond Wiles and the referees, had access to the manuscript. There was a growing clamour for more openness, so everyone could see for themselves the details of the error. The hope was that somebody somewhere might see something that Wiles had missed, and conjure up a calculation to fix the gap in the proof. Some mathematicians claimed that the proof was too valuable to be left in the hands of just one man. Number theorists had become the butt of jibes from other mathematicians, who sarcastically questioned whether or not they understood the concept of proof. What should have been the proudest moment in the history of mathematics was turning into a joke.

Despite the pressure Wiles refused to release the manuscript. After seven years of devoted effort he was not ready to sit back and watch someone else complete the proof and steal the glory. The person who proves Fermat's Last Theorem is not the person that puts in the most work, it's the person who delivers the final and complete proof. Wiles knew that once the manuscript was published in its flawed state he would immediately be swamped by questions and demands for clarification from would-be gap-fixers, and these distractions would destroy his own hopes of mending the proof while giving others vital clues.

Wiles attempted to return to the same state of isolation which had allowed him to create the original proof, and reverted to his habit of studying intensely in his attic. Occasionally he would wander down by the Princeton lake, as he had done in the past. The joggers, cyclists and rowers who had previously passed him by with a brief wave now stopped and asked him whether there was any progress with the gap. Wiles had appeared on front pages around the world, he had been featured in *People* magazine and he had even been interviewed on CNN. The previous summer Wiles had become the world's first mathematical celebrity, and already his image was tarnished.

Meanwhile back in the mathematics department the gossip continued. Princeton mathematician Professor John H. Conway remembers the atmosphere in the department's tea-room: 'We'd gather for tea at 3 o'clock and make a rush for the cookies. Sometimes we'd discuss mathematical problems, sometimes we'd discuss the O.J. Simpson trial, and sometimes we'd discuss

Andrew's progress. Because nobody actually liked to come out and ask him how he's getting on with the proof, we were behaving a little bit like Kremlinologists. So somebody would say: "I saw Andrew this morning" – "Did he smile?" – "Well, yes, but he didn't look too happy." We could only gauge his feelings by his face.'

The Nightmare E-mail

As winter deepened, hopes of a breakthrough faded, and more mathematicians argued that it was Wiles's duty to release the manuscript. The rumours continued and one newspaper article claimed that Wiles had given up and that the proof had irrevocably collapsed. Although this was an exaggeration, it was certainly true that Wiles had exhausted dozens of approaches which might have circumvented the error and he could see no other potential routes to a solution.

Wiles admitted to Peter Sarnak that the situation was getting desperate and that he was on the point of accepting defeat. Sarnak suggested that part of the difficulty was that Wiles had nobody he could trust on a day-to-day basis; there was nobody he could bounce ideas off or who could inspire him to explore more lateral approaches. He suggested that Wiles took somebody into his confidence and try once more to fill the gap. Wiles needed somebody who was an expert in manipulating the Kolyvagin–Flach method and who could also keep the details of the problem secret. After giving the matter prolonged thought, he decided to invite Richard Taylor, a Cambridge lecturer, to Princeton to work alongside him.

Taylor was one of the referees responsible for verifying the proof and a former student of Wiles, and as such he could be doubly trusted. The previous year he had been in the audience at the

Richard Taylor

Isaac Newton Institute watching his former supervisor present the proof of the century. Now it was his job to help rescue the flawed proof.

By January Wiles, with the help of Taylor, was once again tirelessly exploring the Kolyvagin–Flach method, trying to find a way out of the problem. Occasionally after days of effort they would enter new territory, but inevitably they would find themselves back where they started. Having ventured further than ever before and failing over and over again, they both realised that they were in the heart of an unimaginably vast labyrinth. Their deepest fear was that the labyrinth was infinite and without exit, and that they would be doomed to wander aimlessly and endlessly.

Then in the spring of 1994, just when it looked as though things could not get any worse, the following e-mail hit computer screens around the world:

```
Date :       03 April 94
SUBJECT:     Fermat again!

There has been a really amazing development today
on Fermat's Last Theorem.

Noam Elkies has announced a counter-example, so
that Fermat's Last Theorem is not true after all!
He spoke about this at the Institute today. The
solution to Fermat that he constructs involves an
incredibly large prime exponent (larger than
10^20), but it is constructive. The main idea
seems to be a kind of Heegner point construction,
combined with a really ingenious descent for
passing from the modular curves to the Fermat
curve. The really difficult part of the argument
seems to be to show that the field of definition of
```

the solution (which, a priori, is some ring class
field of an imaginary quadratic field) actually
descends to Q.

I wasn't able to get all the details, which were
quite intricate ...

So it seems that the Taniyama–Shimura conjecture
is not true after all. The experts think that it
can still be salvaged, by extending the concept
of automorphic representation, and introducing a
notion of "anomalous curves" that would still
give rise to a "quasi-automorphic
representation".

Henri Darmon
Princeton University

Noam Elkies was the Harvard professor who back in 1988 had
found a counter-example to Euler's conjecture, thereby proving
that it was false:

$$2{,}682{,}440^4 + 15{,}365{,}639^4 + 18{,}796{,}760^4 = 20{,}615{,}673^4.$$

Now he had apparently discovered a counter-example to Fermat's
Last Theorem, proving that it too was false. This was a tragic blow
for Wiles – the reason he could not fix the proof was that the so-
called error was a direct result of the falsity of the Last Theorem. It
was an even greater blow for the mathematical community at large,
because if Fermat's Last Theorem was false, then Frey had already
shown that this would lead to an elliptic equation which was *not*
modular, a direct contradiction to the Taniyama–Shimura conjecture. Elkies had not only found a counter-example to Fermat, he
had indirectly found a counter-example to Taniyama–Shimura.

The death of the Taniyama–Shimura conjecture would have devastating repercussions throughout number theory, because for two decades mathematicians had tacitly assumed its truth. In Chapter 5 it was explained that mathematicians had written dozens of proofs which began with 'Assuming the Taniyama–Shimura conjecture', but now Elkies had shown that this assumption was wrong and all those proofs had simultaneously collapsed. Mathematicians immediately began to demand more information and bombarded Elkies with questions, but there was no response and no explanation as to why he was remaining tight-lipped. Nobody could even find the exact details of the counter-example.

After one or two days of turmoil some mathematicians took a second look at the e-mail and began to realise that, although it was typically dated 2 April or 3 April, this was a result of having received it second or third hand. The original message was dated 1 April. The e-mail was a mischievous hoax perpetrated by the Canadian number theorist Henri Darmon. The rogue e-mail served as a suitable lesson for the Fermat rumour-mongers, and for a while the Last Theorem, Wiles, Taylor and the damaged proof were left in peace.

That summer Wiles and Taylor made no progress. After eight years of unbroken effort and a lifetime's obsession Wiles was prepared to admit defeat. He told Taylor that he could see no point in continuing with their attempts to fix the proof. Taylor had already planned to spend September in Princeton before returning to Cambridge, and so despite Wiles's despondency, he suggested they persevere for one more month. If there was no sign of a fix by the end of September, then they would give up, publicly acknowledge their failure and publish the flawed proof to allow others an opportunity to examine it.

The Birthday Present

Although Wiles's battle with the world's hardest mathematical problem seemed doomed to end in failure, he could look back at the last seven years and be reassured by the knowledge that the bulk of his work was still valid. To begin with Wiles's use of Galois groups had given everybody a new insight into the problem. He had shown that the first element of every elliptic equation could be paired with the first element of a modular form. Then the challenge was to show that if one element of the elliptic equation was modular, then so must the next piece be modular, and so must they all be modular.

During the middle years Wiles wrestled with the concept of extending the proof. He was trying to complete an inductive approach and had wrestled with Iwasawa theory in the hope that this would demonstrate that if one domino fell then they all would. Initially Iwasawa theory seemed powerful enough to cause the required domino effect but in the end it could not quite live up to his expectation. He had devoted two years of effort to a mathematical dead end.

In the summer of 1991, after a year in the doldrums, Wiles encountered the method of Kolyvagin and Flach and he abandoned Iwasawa theory in favour of this new technique. The following year the proof was announced in Cambridge and he was proclaimed a hero. Within two months the Kolyvagin–Flach method was shown to be flawed, and ever since the situation had only worsened. Every attempt to fix Kolyvagin–Flach had failed.

All of Wiles's work apart from the final stage involving the Kolyvagin–Flach method was still worthwhile. The Taniyama–Shimura conjecture and Fermat's Last Theorem might not have

been solved; nevertheless he had provided mathematicians with a
whole series of new techniques and strategies which they could
exploit to prove other theorems. There was no shame in Wiles's
failure and he was beginning to come to terms with the prospect of
being beaten.

As a consolation he at least wanted to understand why he had
failed. While Taylor re-explored and re-examined alternative
methods, Wiles decided to spend September looking one last time
at the structure of the Kolyvagin–Flach method to try and pinpoint
exactly why it was not working. He vividly remembers those final
fateful days: 'I was sitting at my desk one Monday morning, 19
September, examining the Kolyvagin–Flach method. It wasn't
that I believed I could make it work, but I thought that at least I
could explain why it didn't work. I thought I was clutching at
straws, but I wanted to reassure myself. Suddenly, totally unex-
pectedly, I had this incredible revelation. I realised that, although
the Kolyvagin–Flach method wasn't working completely, it was all
I needed to make my original Iwasawa theory work. I realised that
I had enough from the Kolyvagin–Flach method to make my original approach to the problem from three years earlier work. So out
of the ashes of Kolyvagin–Flach seemed to rise the true answer to
the problem.'

Iwasawa theory on its own had been inadequate. The
Kolyvagin–Flach method on its own was also inadequate.
Together they complemented each other perfectly. It was a
moment of inspiration that Wiles will never forget. As he
recounted these moments, the memory was so powerful that he
was moved to tears: 'It was so indescribably beautiful; it was so
simple and so elegant. I couldn't understand how I'd missed it and
I just stared at it in disbelief for twenty minutes. Then during the
day I walked around the department, and I'd keep coming back to

my desk looking to see if it was still there. It was still there. I couldn't contain myself, I was so excited. It was the most important moment of my working life. Nothing I ever do again will mean as much.'

This was not only the fulfilment of a childhood dream and the culmination of eight years of concerted effort, but having been pushed to the brink of submission Wiles had fought back to prove his genius to the world. The last fourteen months had been the most painful, humiliating and depressing period of his mathematical career. Now one brilliant insight had brought an end to his suffering.

'So the first night I went back home and slept on it. I checked through it again the next morning and by 11 o'clock I was satisfied, and I went down and told my wife, "I've got it! I think I've found it." And it was so unexpected that she thought I was talking about a children's toy or something, and she said, "Got what?" I said, "I've fixed my proof. I've got it."'

The following month Wiles was able to make up for the promise he had failed to keep the previous year. 'It was coming up to Nada's birthday again and I remembered that last time I could not give her the present she wanted. This time, half a minute late for our dinner on the night of her birthday, I was able to give her the complete manuscript. I think she liked that present better than any other I had ever given her.'

Subject: Update on Fermat's Last Theorem
Date: 25 Oct 1994 11:04:11

As of this morning, two manuscripts have been released:

Modular elliptic curves and Fermat's Last Theorem,

by Andrew Wiles.

Ring theoretic properties of certain Hecke algebras,

by Richard Taylor and Andrew Wiles.

The first one (long) announces a proof of, among other things, Fermat's Last Theorem, relying on the second one (short) for one crucial step.

As most of you know, the argument described by Wiles in his Cambridge lectures turned out to have a serious gap, namely the construction of an Euler system. After trying unsuccessfully to repair that construction, Wiles went back to a different approach, which he had tried earlier but abandoned in favor of the Euler system idea. He was able to complete his proof, under the hypothesis that certain Hecke algebras are local complete intersections. This and the rest of the ideas described in Wiles' Cambridge lectures are written up in the first manuscript. Jointly, Taylor and Wiles establish the necessary property of the Hecke algebras in the second paper.

The overall outline of the argument is similar to the one Wiles described in Cambridge. The new approach turns out to be significantly simpler and shorter than the original one, because of the removal of the Euler system. (In fact, after seeing these manuscripts Faltings has apparently come up with a further significant simplification of that part of the argument.)

Versions of these manuscripts have been in the hands of a small number of people for (in some cases) a few weeks. While it is wise to be cautious for a little while longer, there is certainly reason for optimism.

Karl Rubin
Ohio State University

Chapter 1

This chapter is devoted to the study of certain Galois representations. In the first section we introduce and study Mazur's deformation theory and discuss various refinements of it. These refinements will be needed later to make precise the correspondence between the universal deformation rings and the Hecke rings in Chapter 2. The main results needed are Proposition 1.2 which is used to interpret various generalized cotangent spaces as Selmer groups and (1.7) which later will be used to study them. At the end of the section we relate these Selmer groups to ones used in the Bloch-Kato conjecture, but this connection is not needed for the proofs of our main results.

In the second section we extract from the results of Poitou and Tate on Galois cohomology certain general relations between Selmer groups as Σ varies, as well as between Selmer groups and their duals. The most important observation of the third section is Lemma 1.10(i) which guarantees the existence of the special primes used in Chapter 3 and [TW].

1. Deformations of Galois representations

Let p be an odd prime. Let Σ be a finite set of primes including p and let $\mathbf{Q}_\Sigma$ be the maximal extension of $\mathbf{Q}$ unramified outside this set and ∞. Throughout we fix an embedding of $\overline{\mathbf{Q}}$, and so also of $\mathbf{Q}_\Sigma$, in $\mathbf{C}$. We will also fix a choice of decomposition group D_q for all primes q in $\mathbf{Z}$. Suppose that k is a finite field of characteristic p and that

$$(1.1) \qquad \rho_0 \colon \mathrm{Gal}(\mathbf{Q}_\Sigma/\mathbf{Q}) \to \mathrm{GL}_2(k)$$

is an irreducible representation. In contrast to the introduction we will assume in the rest of the paper that ρ_0 comes with its field of definition k. Suppose further that $\det \rho_0$ is odd. In particular this implies that the smallest field of definition for ρ_0 is given by the field k_0 generated by the traces but we will not assume that $k = k_0$. It also implies that ρ_0 is absolutely irreducible. We consider the deformations $[\rho]$ to $\mathrm{GL}_2(A)$ of ρ_0 in the sense of Mazur [Ma1]. Thus if $W(k)$ is the ring of Witt vectors of k, A is to be a complete Noetherian local $W(k)$-algebra with residue field k and maximal ideal m, and a deformation $[\rho]$ is just a strict equivalence class of homomorphisms $\rho \colon \mathrm{Gal}(\mathbf{Q}_\Sigma/\mathbf{Q}) \to \mathrm{GL}_2(A)$ such that $\rho \bmod m = \rho_0$, two such homomorphisms being called strictly equivalent if one can be brought to the other by conjugation by an element of $\ker : \mathrm{GL}_2(A) \to \mathrm{GL}_2(k)$. We often simply write ρ instead of $[\rho]$ for the equivalence class.

The first page of Wiles's published proof, which goes on for over a hundred pages.

Andrew Wiles

8

Grand Unified Mathematics

A reckless young fellow from Burma,
Found proofs of the theorem of Fermat,
He lived then in terror,
Of finding an error,
Wiles' proof, he suspected, was firmer!

Fernando Gouvea

This time there was no doubt about the proof. The two papers, consisting of 130 pages in total, were the most thoroughly scrutinised mathematical manuscripts in history and were eventually published in *Annals of Mathematics* (May, 1995).

Once again Wiles found himself on the front page of the *New York Times*, but this time the headline 'Mathematician Calls Classic Riddle Solved' was overshadowed by another science story – 'Finding on Universe's Age Poses New Cosmic Puzzle'. While journalists were slightly less enthusiastic about Fermat's Last Theorem this time around, the mathematicians had not lost sight of the true significance of the proof. 'In mathematical terms the final proof is the equivalent of splitting the atom or finding the structure of DNA,' announced John Coates. 'A proof of Fermat is a great intellectual triumph and one shouldn't lose sight of the fact that it has revolutionised number theory in one fell swoop. For me the charm and beauty of Andrew's work has been that it has been a tremendous step for algebraic number theory.'

During Wiles's eight-year ordeal he had brought together virtually all the breakthroughs in twentieth-century number theory and incorporated them in one almighty proof. He had created completely new mathematical techniques and combined them with traditional ones in ways that had never been considered possible. In doing so he had opened up new lines of attack on a whole host of other problems. According to Ken Ribet the proof is a perfect synthesis of modern mathematics and an inspiration for the future: 'I think that if you were lost on a desert island and you had only this manuscript then you would have a lot of food for thought. You would see all of the current ideas of number theory. You turn to a page and there's a brief appearance of some fundamental theorem by Deligne and then you turn to another page and in some incidental way there's a theorem by Hellegouarch – all of these things are just called into play and used for a moment before going on to the next idea.'

While science journalists eulogised over Wiles's proof of Fermat's Last Theorem, few of them commented on the proof of the Taniyama–Shimura conjecture which was inextricably linked to it. Few of them bothered to mention the contribution of Yutaka Taniyama and Goro Shimura, the two Japanese mathematicians who back in the 1950s had sown the seeds for Wiles's work. Although Taniyama had committed suicide over thirty years ago, his colleague Shimura was there to witness their conjecture proved. When asked for his reaction to the proof, Shimura gently smiled and in a restrained and dignified manner simply said, 'I told you so.'

Like many of his colleagues, Ken Ribet feels that proving the Taniyama–Shimura conjecture has transformed mathematics: 'There's an important psychological repercussion which is that people now are able to forge ahead on other problems that they were too timid to work on before. The landscape is different, in that you know that all elliptic equations are modular and therefore when you prove a theorem for elliptic equations you're also attacking modular forms and vice versa. You have a different perspective of what's going on and you feel less intimidated by the idea of working with modular forms because basically you're now working with elliptic equations. And, of course, when you write an article about elliptic equations, instead of saying that we don't know anything so we're going to have to assume the Taniyama–Shimura conjecture and see what we can do with it, now we can just say that we know the Taniyama–Shimura conjecture is true, so therefore such and such must be true. It's a much more pleasant experience.'

Via the Taniyama–Shimura conjecture Wiles had unified the elliptic and modular worlds, and in so doing provided mathematics with a short cut to many other proofs – problems in one domain could be solved by analogy with problems in the parallel domain. Classic unsolved elliptic problems dating all the way back to the ancient Greeks could now be re-examined using all the available modular tools and techniques.

Even more importantly, Wiles had made the first step towards Robert Langlands's grander scheme of unification – the Langlands programme. There is now a renewed effort to prove other unifying conjectures between other areas of mathematics. In March 1996 Wiles shared the $100,000 Wolf Prize (not to be confused with the Wolfskehl Prize) with Langlands. The Wolf Committee was recognising that while Wiles's proof was an astounding accomplishment in its own right, it had also breathed life into Langlands's ambitious scheme. Here was a breakthrough which could lead mathematics into the next golden age of problem-solving.

Following a year of embarrassment and uncertainty the mathematical community could at last rejoice. Every symposium, colloquium and conference had a session devoted to Wiles's proof and in Boston mathematicians launched a limerick competition to commemorate the momentous event. It attracted this entry:

> 'My butter, garçon, is writ large in!'
> A diner was heard to be chargin',
> 'I had to write there',
> Exclaimed waiter Pierre,
> 'I couldn't find room in the margerine.'

> E. Howe, H. Lenstra, D. Moulton

Great Unsolved Problems

Wiles realises that in order to give mathematics one of its greatest proofs, he has had to deprive it of its greatest riddle: 'People have told me that I've taken away their problem, and asked if I could give them something else. There is a sense of melancholy. We've lost something that's been with us for so long, and something that drew a lot of us into mathematics. Perhaps that's always the way with maths problems. We just have to find new ones to capture our attention.'

Even though Wiles has now disposed of the most famous problem in mathematics, puzzlers around the world need not lose hope, because there is still a multitude of unsolved mathematical conundrums. Many of these profound problems, like Fermat's Last Theorem, have their roots in the mathematics of ancient Greece and can be understood by a schoolchild. For example, there are still mysteries concerning the perfect numbers. As discussed in

Chapter 1, perfect numbers are those numbers whose divisors add up to the number itself. For example, 6 and 28 are perfect numbers because

$$1, 2, 3 \text{ divide into } 6 \quad \text{and} \quad 6 = 1 + 2 + 3,$$

$$1, 2, 4, 7, 14 \text{ divide into } 28 \quad \text{and} \quad 28 = 1 + 2 + 4 + 7 + 14.$$

René Descartes said that 'perfect numbers like perfect men are very rare', and indeed over the last few thousand years only thirty have been discovered. The newest and largest perfect number contains 130,000 digits and is defined by the formula

$$2^{216,090} \times (2^{216,091} - 1).$$

One thing which all the known perfect numbers have in common is that they are even, which might suggest that all perfect numbers are even. An obvious, and as it turns out frustrating, challenge would be to prove that this is true – are all perfect numbers even?

The other great puzzle about perfect numbers asks if there is an inexhaustible supply of them. Through the centuries thousands of number theorists have tried and failed to prove that there is, or is not, an infinite number of perfect numbers. Whoever succeeds will automatically earn a place in history.

Another area of mathematics rich in ancient unsolved problems is the theory of prime numbers. The sequence of prime numbers forms no discernible pattern and as such the primes are a law unto themselves. They have been described as weeds growing randomly among the counting numbers. While checking through the natural numbers it is possible to find regions rich in primes, but for no known reason other regions are completely barren. For centuries mathematicians have tried and failed to explain the underlying pattern behind the primes. It could be that no pattern exists, and that the primes exhibit an inherently random distribution, in which case mathematicians would be advised to tackle other less ambitious prime problems.

For example, two thousand years ago Euclid proved that there was an inexhaustible supply of prime numbers (see Chapter 2), but for the last two centuries mathematicians have been trying to prove that there is also an inexhaustible supply of twin primes. Twin primes are pairs of primes which differ by only 2, which is as close as primes can be to each other – they cannot differ by 1 because then one of them would have to be even and therefore divisible by 2, and therefore not a prime. Examples of small twin primes are (5, 7) and (17, 19), and larger ones include (22,271, 22,273) and (1,000,000,000,061, 1,000,000,000,063). Twin primes seem to be sprinkled throughout the sequence of whole numbers, and the harder mathematicians search for them, the more they find. There is strong evidence that there is an infinite number of them but nobody has ever been able to prove that this is the case.

The most recent breakthrough towards proving the so-called *twin prime conjecture* was made back in 1966 when the Chinese mathematician Chen Jing-run was able to show there are an infinite number of prime and *almost prime* pairs. Actual primes have no factors, other than 1 and the number itself, but almost primes are the next best thing, because they have only two prime factors. So 17 is a prime number, but 21 (3×7) is almost prime. Numbers such as 120 ($2 \times 3 \times 4 \times 5$) are not prime at all because they are the product of several prime factors. Chen proved that there were an infinite number of cases where a prime number was twinned with either another prime number or an almost prime number. Whoever can go one step further and remove the 'almost' will have achieved the biggest breakthrough in prime number theory since Euclid.

Another prime number riddle dates back to 1742 when Christian Goldbach, tutor to the teenage Czar Peter II, wrote a letter to the great Swiss mathematician Leonhard Euler. Goldbach had examined dozens of even numbers and noticed that he could split all of them into the sum of two primes:

$$4 = 2 + 2,$$
$$6 = 3 + 3,$$
$$8 = 3 + 5,$$
$$10 = 5 + 5,$$
$$50 = 19 + 31,$$
$$100 = 53 + 47,$$
$$21{,}000 = 17 + 20{,}983,$$
$$\vdots$$

Goldbach asked Euler if he could prove that every even number could be split into two primes. Despite years of effort, the man known as 'analysis incarnate' was confounded by Goldbach's challenge. In today's computer age the Goldbach conjecture, as it has become known, has been tested and found to be correct for every even number up to 100,000,000 but still nobody has been able to prove that the conjecture holds true for every even number up to infinity. Mathematicians have been able to prove that every even number is the sum of no more than 800,000 primes, but this is a long way from proving the original conjecture. Even so, such weaker proofs have yielded important insights into the nature of primes and in 1941 Stalin awarded a prize of 100,000 Roubles to the Russian mathematician Ivan Matveyevich Vinogradov, who had gone some way towards proving the Goldbach conjecture.

Of all the problems likely to replace Fermat's Last Theorem as the greatest unsolved problem in mathematics the best candidate is Kepler's sphere-packing problem. In 1609 the German scientist Johannes Kepler showed that the planets moved in elliptical rather than circular orbits, a discovery which revolutionised astronomy and which would later inspire Isaac Newton to deduce the law of universal gravitation. Kepler's mathematical legacy is somewhat less grand in scale but equally profound. It essentially concerns the curious problem of arranging piles of oranges in the most efficient manner.

The problem was born in 1611 when Kepler wrote a paper entitled 'On the six-cornered snowflake', a New Year's gift for his patron John Wacker of Wackenfels. He successfully explained why all snowflakes have a unique but always hexagonal structure, by suggesting that every snowflake begins with a hexagonally symmetric seed which grows as it falls through the atmosphere. Continually changing conditions of wind, temperature and moisture ensure that each snowflake is unique, and yet the seed is so small that the conditions determining the pattern of growth will be identical on all of the six sides, ensuring that symmetry is maintained. In this apparently light-hearted paper Kepler, who had a remarkable talent for drawing deep insights from the simplest observations, was laying the foundations of crystallography.

Kepler's interest in how particles of matter arrange and apparently organise themselves led him to discuss another question, namely what is the most efficient way to stack particles such that they occupy the least possible volume? If the particles are assumed to be spheres it is clear that, however they are arranged, there will inevitably be gaps between them, and the challenge was to identify which arrangement would minimise the gaps. In order to solve the problem Kepler constructed various arrangements and then calculated the packing efficiency for each one.

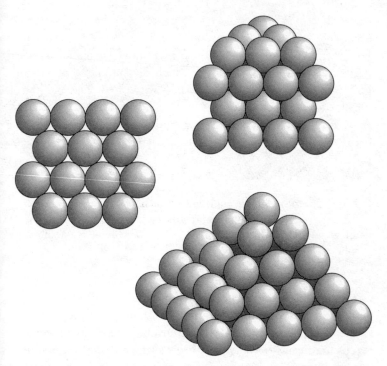

Figure 24. In the face-centred cubic arrangement each layer consists of spheres arranged so that each one is surrounded by six others. One layer is then placed horizontally above another such that a sphere sits in a dimple rather than directly above another sphere. A particular orientation of this arrangement gives rise to the familiar greengrocer's pyramid of oranges.

One of the first arrangements examined by Kepler is now known as the face-centred cubic lattice. This can be constructed by first creating a bottom layer of spheres such that each sphere is

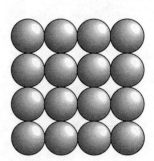

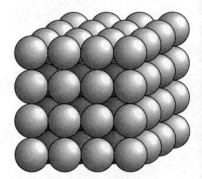

Figure 25. In the simple cubic arrangement each layer consists of spheres arranged in a square grid. One layer is placed horizontally above another such that each sphere sits directly above another sphere.

surrounded by six other spheres. The second layer is generated by positioning spheres in the 'dimples' of the first layer, as shown in Figure 24. The second layer is effectively a duplicate of the first except that it has been shifted across slightly so that it fits snugly into position. This arrangement is identical to the one used by greengrocers stacking pyramids of oranges, and has an efficiency of 74%. This means that if a large cardbox box were to be filled with oranges using the face-centred strategy then the oranges would occupy 74% of the box's volume.

This arrangement can be compared to others such as the simple cubic lattice. In this case each layer consists of spheres positioned in a square grid formation and the layers are placed directly above each other as shown in Figure 25. The simple cubic lattice has a packing efficiency of only 53%.

Another arrangement, the hexagonal lattice, is similar to the face-

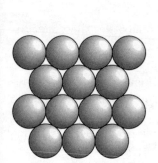 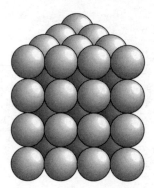

Figure 26. In the hexagonal lattice arrangement each layer consists of spheres arranged so that each one is surrounded by six others. One layer is then placed horizontally above another such that each sphere sits directly above another sphere.

centred cubic in as much as each layer is made up by surrounding every sphere by six others, but instead of slightly displacing each layer so that it sits neatly in the dimples of the one below it, the layers are positioned directly above each other, as shown in Figure 26. The hexagonal lattice has a packing efficiency of only 60%.

Kepler studied a whole variety of configurations and came to a conclusion which he felt worthy of including in his paper 'On the six-cornered snowflake', namely that it was the face-centred cubic lattice for which 'the packing will be the tightest possible'. Kepler's statement was perfectly sensible because the packing efficiency for the face-centred cubic lattic was the best he had found, but this did not rule out the possibility that there was some arrangement with an even higher packing efficiency which he had overlooked. This small element of doubt is at the heart of the sphere-packing problem, a riddle which pre-dates Fermat by half a century and which has now turned out to be even more intractable than the Last Theorem. The problem requires mathematicians to prove that the face-centred cubic lattice is without doubt the most efficient method of packing spheres.

Like the Last Theorem, Kepler's problem requires mathematicians to develop a proof which can encompass an infinity of possibilities. Fermat claimed that among the infinity of whole numbers there are no solutions to his equation, and Kepler claimed that among the infinity of arrangements none would have a packing efficiency higher than that of the face-centred cubic lattice. As well as having to demonstrate that there are no other lattices, that is to say regular arrangements, with a higher packing efficiency, mathematicians would also have to include all possible random arrangements in their proof.

In the last 380 years nobody has been able to prove that the face-centred cubic is indeed the optimum packing strategy; on the other hand nobody has discovered a more efficient packing method. The lack of a counter-example means that for all practical purposes Kepler's statement is effectively true, but in the absolute world of mathematics a rigorous proof is still required. This led the British sphere-packing expert C. A. Rogers to comment that Kepler's claim is one that 'most mathematicians believe, and all physicists know'.

Despite the lack of a complete proof there have over the centuries been a few milestones towards a solution. In 1892 the Scandinavian mathematician Axel Thue provided a proof to the two-dimensional analogue of Kepler's problem, i.e. what is the most efficient way to arrange spheres when considering only a single layer, or in other words arranging oranges in a tray rather than a box. The solution is the hexagonal arrangement. Subsequently Tóth, Segre and Mahler came to the same conclusion, but none of these methods could be applied to the original three-dimensional Kepler problem.

In the modern era some mathematicians have tried a rather different tack, which is to put an upper limit on the possible packing efficiency. In 1958 C. A. Rogers calculated an upper limit of 77.97% – this meant that it is impossible to have an arrangement with a packing efficiency greater than 77.97%. This percentage is not much higher than the packing efficiency of the face-centred cubic lattice, 74.04%. Therefore, if any arrangement were to have a higher efficiency than the face-centred cubic, then it could only beat it by a few per cent. There was only a tiny window of 3.93% by which a rogue arrangement could enter and prove Kepler wrong. After Rogers, other mathematicians began to try and close the window completely by reducing the upper limit to 74.04%, which would then leave no room for any other arrangement to beat the efficiency of the face-centred cubic and therefore prove Kepler right by default. Unfortunately reducing the upper limit has turned out to be a slow and difficult process and by 1988 it stood at 77.84%, only marginally better than Rogers's result.

Despite years of slow progress the sphere-packing problem suddenly hit the headlines in the summer of 1990 when Wu-Yi Hsiang of the University of California at Berkeley published a result which he claimed was a proof of Kepler's conjecture. Initially the mathematical community responded optimistically, but, as with Wiles's proof, the paper had to undergo a process of peer review before it could be accepted as valid. As the weeks passed Hsiang was confronted with a series of blunders and the proof was left in tatters.

In a story which parallels Wiles's ordeal Hsiang responded a year later with a revised proof which he claimed circumvented the problems identified in the original manuscript. Unfortunately for Hsiang his critics still believed that there were gaps in his logic. In a letter to Hsiang the mathematician Thomas Hales tried to explain his doubts:

One assumption made in your second paper strikes me as more fundamental and yet far more difficult to prove than the others . . . You state
that 'the best way (that is volume-minimising) of adding a second layer of
packing is to cap as many holes as possible . . .' Your argument seems to
rely heavily and essentially on this assumption yet nowhere is there even
a hint of a proof.

Since Hsiang's revamped paper there has been an ongoing battle
between him and his critics, with claims and counterclaims that the
problems have and have not been resolved. At best the proof is still
shrouded in controversy, at worst it has been discredited – either
way the door is still open for anybody who wants to prove Kepler's
conjecture. In 1996 Doug Muder gave a personal summary of the
situation, which also revealed some of the intrigue surrounding
Hsiang's proof:

I've recently returned from the AMS-IMS-SIAM Joint Summer
Research Conference on Discrete and Computational Geometry at
Mount Holyoke. This was a once-every-ten-years conference, and so
there was a focus on assessing the progress of the last ten years. Hsiang's
claim to have proved Kepler's conjecture is now six years old, and I found
that the community has reached consensus on it: no one buys it.

During the plenary lectures and the informal discussions in the cafete-
ria, the following points were never in dispute:

1. Hsiang's paper (published in the *International Journal of Mathematics* in
1993) is not a proof of the Kepler conjecture. At best it is a sketch (a 100-
page sketch!) of how such a proof might go.

2. Even as a sketch the paper is inadequate, since counter-examples
have been found to several of its steps.

3. Hsiang's related claim to have proved the Dodecahedron conjecture
(and various other previously unsolved sphere-packing problems) is
equally baseless.

4. Work on the Kepler conjecture and the Dodecahedron conjecture should continue as if Hsiang's paper had never existed.

In one lecture Gabor Fejes Tóth of the Hungarian Academy of Sciences said of Hsiang's paper, 'This cannot be considered a proof. The problem is still open.' Thomas Hales of the University of Michigan agreed: 'This problem is still unsolved. I haven't solved it. Hsiang hasn't solved it. Nobody else has solved it as far as I know.' (Hales did predict that his own techniques would solve the problem 'in a year or two'.)

What makes this an interesting story is that one person is still absent from the consensus – Hsiang himself. (He also did not attend the conference.) He is well aware of the counter-examples and the fact that his claims are not believed by the experts in the field, and yet he continues to give lectures all over the world, repeating his claims. People who have interacted personally (such as Hales and Bezdek) believe that he will never admit that his paper is flawed.

This is the reason why it has taken so long for the dust to settle. Hsiang first claimed a solution to the Kepler conjecture in 1990, six years ago. His talks have consistently been vague enough to be plausible. Many months after the first claims, when the first preprint appeared, gaps were spotted almost immediately, and counter-examples quickly followed. But the fact that Hsiang continued making his claims in public created the impression that he must have dealt with whatever objections had surfaced to date. The length of the paper, and the fact that it went through several iterations prior to publication, added to the confusion.

The case of Hsiang demonstrates to what extent mathematics relies on an honor system. The community assumes that tenured professors at top-flight universities will not make spurious claims, and will retract incorrect claims at the first demonstrated flaw. Someone who flouts this system can create confusion for a very long time, since no one has the time or the motivation to follow him around and debunk his claims whenever he makes them. (When you consider the amount of work that must have gone into Hales' debunking article in 1993 in the *Mathematical Intelligencer* – and the fact that this article does nothing to further his research career – you begin to see the problem. Hsiang's published rejoinder to that article was completely inadequate, but Hales concluded that debunking Hsiang's rejoinder would start a never-ending cycle that he simply did not have time for.)

Hsiang may never admit his mistakes, but what about the *International Journal*? Clearly they are part of the process that did not work the way it is supposed to. Hsiang's paper was not adequately refereed, if it was refereed at all. The fact that the *Journal* is edited by Hsiang's Berkeley colleagues lends an air of cronyism to the story. The *Journal* had no interest in sphere-packing until this paper. It seems clear that Hsiang chose the *International Journal* because it was edited by his friends, not because it was the appropriate venue for his paper.

Karoly Bezdek, who spent more than a year working with Hsiang in an attempt to fill the gaps in the paper, has submitted a paper to the *Journal* containing a counter-example to one of Hsiang's lemmas. They have been sitting on it since December – not an unusual time for a paper to be refereed, but quite long for a counter-example to the *Journal*'s most publicised paper in many years.

Doug Muder

Silicon Proofs

In his battle against Fermat's Last Theorem, Wiles's only weapons were a pencil, paper and pure logic. Although his proof employs the most modern techniques in number theory it is in the best tradition of Pythagoras and Euclid. However, recently there have been ominous signs that Wiles's solution may be one of the last examples of a heroic proof and that future results may rely on a

Francis Guthrie realised that he could colour a county map of
Britain with only four colours and, at the same time, avoid any
neighbouring counties sharing the same colour. He then wondered
if any conceivable map would require more than four colours.

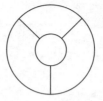

Figure 27. This simple pattern shows that at least four colours are required for some maps, but are four colours sufficient for all maps?

brute force approach rather than an elegant argument.

The first indication of what some are calling the decline of mathematics concerns a problem created in England in October 1852 by the part-time mathematician Francis Guthrie. One afternoon, while idly colouring a map of the counties of Britain, Guthrie stumbled upon a conundrum which appeared trivial but which he could not solve. He simply wanted to know the minimum number of colours required to colour any conceivable map, such that no two regions having a common border have the same colour.

For example, three colours are not enough for the pattern in Figure 27. Therefore it is clear that some maps require four colours, but Guthrie wanted to know if four colours would be enough for all maps, or might some require five, six, or more colours?

Frustrated but intrigued, Guthrie mentioned the problem to his younger brother, Frederick, who was a student at University College, London. In turn Frederick posed the problem to his professor, the eminent Augustus De Morgan, who on 23 October wrote to the great Irish mathematician and physicist William Rowan Hamilton:

A student of mine asked me today to give him a reason for a fact which I did not know was a fact – and do not yet. He says that if a figure be anyhow divided and the compartments differently coloured so that figures with any portion of common boundary line are differently coloured then four colours may be wanted, but not more. I have a case in which four colours are wanted. Query – cannot a necessity for five or more be invented . . . If you retort with some very simple case which makes me out to be a stupid animal, I think I must do as the Sphinx did . . .

Hamilton was unable to invent a map which required five colours but neither could he prove that no such map existed. News of the problem rapidly spread throughout Europe but it robustly resisted attack from all quarters, proving itself to be deceptively difficult. In a fit of pride Hermann Minkowski said that the reason it had not been solved was that only third-rate mathematicians had attempted it, but his own efforts also ended in failure. 'Heaven is angered by my arrogance,' he announced. 'My proof is also defective.'

Despite inventing one of the toughest problems in mathematics, now known as the *four-colour problem*, Francis Guthrie left England and practised as a barrister in South Africa. He did eventually return to mathematics as a professor at Cape Town University, where he tended to spend more time in the botany department than with his mathematical colleagues – his only claim to fame beyond the four-colour problem was having a heather named after him, *Erica guthriei*.

Having survived unsolved for a quarter of a century there was a tremendous amount of optimism in 1879 when the British mathematician Alfred Bray Kempe published a paper in the *American Journal of Mathematics* in which he claimed a solution to Guthrie's riddle. Kempe appeared to prove that every map required at most four colours and the peer review procedure seemed to confirm it.

He was immediately elected a Fellow of the Royal Society and was eventually knighted for his contribution to mathematics.

Then in 1890 Percy John Heawood, a lecturer at Durham University, published a paper which shocked the mathematics establishment. A decade after Kempe had apparently solved the problem, Heawood showed that the so-called proof was fundamentally flawed. The only good news was that, as part of his demolition of Kempe's work, Heawood was able to show that the maximum number of colours required was either four or five and certainly no higher.

Although Kempe, Heawood, and others were unable to solve the four-colour problem their failed efforts contributed enormously to the new and blossoming topic of topology. Unlike geometry, in which the exact shape and size of an object are being studied, topology is merely interested in the essence of the object, its most basic features. For example, when a geometer examines a square the properties of interest are the equal lengths of each side, and the right angle at each corner. When a topologist examines the same object the only property of interest is that the square is a single unbroken line which effectively forms a loop. Therefore a topologist will view a circle as being indistinguishable from a square, because it too consists of a single loop. The mathematician John Kelley whimsically commented, 'A topologist is one who doesn't know the difference between a doughnut and a coffee cup.'

Another way to view the topological equivalence of a square and a circle is by imagining one of the shapes drawn on a rubber sheet. If we start with the square, then the rubber sheet could be stretched, pulled, bent and twisted (but never torn) until the original shape is transformed into a circle. On the other hand the square could never be transformed into a cross, no matter how much the rubber sheet is deformed. Hence a square and a cross are not topologically equivalent. Because of this way of thinking, topology is often called 'rubber-sheet geometry'.

Having abandoned concepts such as length and angle, topologists can only distinguish between objects by resorting to features such as the number of intersections possessed by an object. In this way a figure-of-eight is fundamentally different from a circle because it contains a point where four lines meet, whereas the circle contains no such intersections. No amount of stretching and twisting will transform the figure-of-eight into a circle. Topologists are also interested in three-dimensional objects (and higher) where holes, loops and knots become the fundamental features of interest.

Mathematicians hoped that by viewing maps through the simplifying lens of topology they would grasp the essence of the four-colour problem. The first breakthrough came in 1922 when Philip Franklin ignored the general problem and settled for a proof which showed that any map containing 25 or fewer regions required only four colours. Other mathematicians attempted to build on Franklin's methods, and in 1926 Reynolds extended the proof to maps with 27 regions; in 1940 Winn extended it to 35 regions; and by 1970 Ore and Stemple had reached 39 regions. The problem seemed to mirror the history of Fermat's Last Theorem: slow progress was being made towards infinity. The original conjecture appeared almost certainly to be true, but until a general proof could be found there was always the possibility that a map would be drawn which would prove Guthrie wrong. In fact, in 1975 the mathematical journalist and writer Martin Gardner published a map in the magazine *Scientific American*, which he claimed required five colours. The date of publication was 1 April and Gardner was well aware that although it was difficult to cover the map with just four colours it was not impossible. You might like to demonstrate

Figure 28. On 1 April 1975 Martin Gardner presented this map in his column in *Scientific American*. He claimed that it required five colours, but of course his claim was a hoax.

that this is the case: the map in question is shown in Figure 28.

The slow pace of progress made it increasingly clear that conventional approaches would never bridge the gap between Ore and Stemple's proof for maps of 39 regions or less and any conceivable map which could consist of an infinite number of regions. Then in 1976 two mathematicians at the University of Illinois, Wolfgang Haken and Kenneth Appel, came up with a new technique which would revolutionise the concept of mathematical proof.

Haken and Appel had been studying the work of Heinrich Heesch who had claimed that the infinity of infinitely variable maps could be constructed from some finite number of finite maps and that by studying these building-block maps it might be possible to get a handle on the general problem. The basic maps were the equivalent of the electron, proton and neutron, the fundamental objects from which all else could be constructed. Unfortunately the situation was not as simple as the holy trinity of particles, because Haken and Appel could only reduce the four-colour problem to 1482 building block configurations. If Haken and Appel could prove that these maps were four-colourable, then this would imply that all maps would be four-colourable.

Checking the 1482 maps and all the colouring combinations within each map would be an immense task, certainly beyond the capability of any team of mathematicians. Even employing a computer to crank through the permutations could take a century. Undaunted, Haken and Appel began to look for short cuts and strategies which a computer could employ to accelerate the map-checking procedure. In 1975, five years after they began working on the problem, the two men witnessed how the computer was doing more than mere calculations, it was contributing to their ideas. The two men recall the pivotal point in their research:

At this point the programme began to surprise us. At the beginning we would check its arguments by hand so we could always predict the course it would follow in any situation; but now it suddenly started to act like a chess-playing machine. It would work out compound strategies based on all the tricks it had been 'taught', and often these approaches were far more clever than those we would have tried. Thus it began to teach us things about how to proceed that we never expected. In a sense it had surpassed its creators in some aspects of the 'intellectual' as well as the mechanical parts of the task.

In June 1976, thanks to 1200 hours of computer time, Haken and Appel were able to announce that all 1482 maps had been analysed and none of them required more than four colours. Guthrie's four-colour problem had at last been solved. What was remarkable was that this was the first mathematical proof in which a computer had done more than just speed up the calculation – it had contributed so much to the result that the proof would have been impossible without it. It was a tremendous achievement but at the same time there was an uneasy feeling in the community because there was no way to check the proof in the conventional sense.

Before the details of the proof could be published in the *Illinois Journal of Mathematics*, the editors had to have some level of peer review. Conventional refereeing was impossible, so instead Haken and Appel's program was fed into an independent computer to demonstrate that it too would achieve the same result.

This unorthodox refereeing process infuriated some mathematicians who claimed that it was an inadequate check, and that there was no guarantee against some glitch in the heart of the computer generating an error in the logic. H.P.F. Swinnerton-Dyer pointed out the following about computer proofs:

When a theorem has been proved with the help of a computer, it is impossible to give an exposition of the proof which meets the traditional test – that a sufficiently patient reader should be able to work through the proof and verify that it is correct. Even if one were to print all the programs and all the sets of data used there can be no assurance that a data tape has not been mispunched or misread. Moreover, every modern computer has obscure faults in its software and hardware – which so seldom cause errors that they go undetected for years – and every computer is liable to transient faults.

To some extent this was paranoia from a community that prefers to shun computers rather than exploit them. Joseph Keller once noted that at his university, Stanford, the mathematics department had fewer computers than any other department, including French Literature. Those mathematicians who rejected the work of Haken and Appel could not deny that all mathematicians accept proofs even if they personally have not checked them. In the case of Wiles's proof of Fermat's Last Theorem, less than 10% of number theorists fully understand the logic but 100% accept it as being true. Those who cannot grasp the proof are satisfied because others who do understand its concepts have examined and verified them.

An even more extreme case is the so-called proof of the classification of finite simple groups, which consists of 500 separate articles written by over a hundred mathematicians. It is said that only one mathematician, Daniel Gorenstein, understood the entire 15,000-page proof, and he died in 1992. However, the community at large can rest assured that every section of the proof has been examined by its own team of specialists and every single line of the 15,000 pages has been checked and double-checked dozens of times. What makes the four-colour theorem different is that it has never been fully checked by anybody and it never will be.

In the twenty years since the proof of the four-colour theorem was announced computers have been used to solve other less famous but equally important problems. In a subject previously untainted by technology more and more mathematicians are reluctantly coming to terms with the increasing use of silicon logic and accepting Wolfgang Haken's argument:

Anyone, anywhere along the line, can fill in the details and check them. The fact that a computer can run through more details in a few hours than a human could ever hope to do in a lifetime does not change the basic concept of mathematical proof. What has changed is not the theory but the practice of mathematics.

Most recently some mathematicians have handed even more power to the computer by employing so-called genetic algorithms. These are computer programs whose broad structure is designed by a mathematician but whose fine detail is determined by the computer itself. Certain lines within the program are allowed to mutate and evolve rather like individual genes in organic DNA. From the original mother program the computer will generate hundreds of daughter programs which are all slightly different because of random mutations made by the computer. The daughter programs are then used to try and solve a particular problem. Most of the programs will fail dismally, but the one which gets furthest towards a result will be allowed to reproduce and create a new generation of mutated daughters. Survival of the fittest is interpreted in terms of which program gets closest to solving a problem. By repeating the process mathematicians hope that without intervention a program will evolve to solve the problem, and in some cases this approach is having significant success.

The computer scientist Edward Frenkin has gone as far as saying that a computer will one day discover an important proof independently of mathematicians. A decade ago he instituted the Leibniz Prize, $100,000 to be awarded to the first computer program to devise a theorem that has a 'profound effect on mathematics'. Whether or not the prize will ever be claimed is a matter of debate, but what is certain is that a computer proof will always lack the illumination of traditional proofs and seem hollow by comparison. A mathematical proof should not only answer a question, it should also give some understanding as to why the answer is what it is. Sending a question into a black box and receiving an answer out of the other end adds to knowledge but not to understanding. In Wiles's proof of Fermat's Last Theorem we know that there are no solutions to Fermat's equation because such a solution would lead to a contradiction with the Taniyama–Shimura conjecture. Not only has Wiles met Fermat's challenge, but he has justified his answer by saying it must be so in order to maintain a fundamental relationship between elliptic equations and modular forms.

The mathematician Ronald Graham described the shallowness of computer proofs in the context of one of today's great unproved conjectures, the Riemann hypothesis: 'It would be very discouraging if somewhere down the line you could ask a computer if the Riemann hypothesis is correct and it said, "Yes, it is true, but you won't be able to understand the proof."' The mathematician Philip Davis, writing with Reuben Hersh, had a similar reaction to the proof of the four-colour problem:

My first reaction was, 'Wonderful! How did they do it?' I expected some brilliant new insight, a proof which had in its kernel an idea whose beauty would transform my day. But when I received the answer, 'They did it by breaking it down into thousands of cases, and then running them all on the computer, one after the other', I felt disheartened. My reactions was, 'So it just goes to show, it wasn't a good problem after all.'

The Prize

Wiles's proof of Fermat's Last Theorem relies on verifying a conjecture born in the 1950s. The argument exploits a series of mathematical techniques developed in the last decade, some of which were invented by Wiles himself. The proof is a masterpiece of modern mathematics, which leads to the inevitable conclusion that Wiles's proof of the Last Theorem is not the same as Fermat's. Fermat wrote that his proof would not fit into the margin of his copy of Diophantus' *Arithmetica*, and Wiles's 100 pages of dense mathematics certainly fulfils this criterion, but surely the Frenchman did not invent modular forms, the Taniyama–Shimura conjecture, Galois groups and the Kolyvagin–Flach method centuries before anyone else.

If Fermat did not have Wiles's proof, then what did he have? Mathematicians are divided into two camps. The hard-headed sceptics believe that Fermat's Last Theorem was the result of a rare moment of weakness by the seventeenth-century genius. They claim that, although Fermat wrote 'I have discovered a truly marvellous proof', he had in fact only found a flawed proof. The exact nature of this flawed proof is open to debate, but it is quite possible that it may have been along the same lines as the work of Cauchy or Lamé.

Other mathematicians, the romantic optimists, believe that Fermat may have had a genuine proof. Whatever this proof might have been, it would have been based on seventeenth-century techniques, and would have involved an argument so cunning that it has eluded everybody from Euler to Wiles. Despite the publication of Wiles's solution to the problem, there are plenty of mathematicians who believe that they can still achieve fame and glory by discovering Fermat's original proof.

Although Wiles had to resort to twentieth-century methods to prove a seventeenth-century riddle, he has nonetheless met Fermat's challenge according to the rules of the Wolfskehl committee. On 27 June 1997 Andrew Wiles collected the Wolfskehl Prize worth $50,000. Once again Fermat and Wiles made headlines around the world. Fermat's Last Theorem had been officially solved.

But what next will capture Wiles's attention? Not surprisingly for a man who worked in complete secrecy for seven years, he is refusing to comment on his current research, but whatever he is working on, there is no doubt that it will never fully replace the infatuation he had with Fermat's Last Theorem. 'There's no other problem that will mean the same to me. This was my childhood passion. There's nothing to replace that. I've solved it. I'll try other problems, I'm sure. Some of them will be very hard and I'll have a sense of achievement again, but there's no other problem in mathematics that could hold me the way Fermat did.

'I had this very rare privilege of being able to pursue in my adult life what had been my childhood dream. I know it's a rare privilege, but if you can tackle something in adult life that means that much to you, then it's more rewarding than anything imaginable. Having solved this problem there's certainly a sense of loss, but at the same time there is this tremendous sense of freedom. I was so obsessed by this problem that for eight years I was thinking about it all the time – when I woke up in the morning to when I went to sleep at night. That's a long time to think about one thing. That particular odyssey is now over. My mind is at rest.'

Appendices

Appendix 1. The Proof of Pythagoras' Theorem

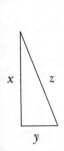

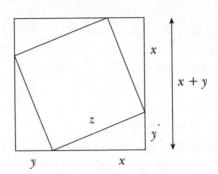

The aim of the proof is to show that Pythagoras' theorem is true for all right-angled triangles. The triangle shown above could be any right-angled triangle because its lengths are unspecified, and represented by the letters x, y and z.

Also above, four identical right-angled triangles are combined with one tilted square to build a large square. It is the area of this large square which is the key to the proof.

The area of the large square can be calculated in two ways.

Method 1: Measure the area of the large square as a whole. The length of each side is $x + y$. Therefore, the area of the large square $= (x + y)^2$.

Method 2: Measure the area of each element of the large square. The area of each triangle is $\frac{1}{2}xy$, i.e. $\frac{1}{2}$ × base × height. The area of the tilted square is z^2. Therefore,

$$\text{area of large square} = 4 \times (\text{area of each triangle}) + \text{area of tilted square}$$
$$= 4\left(\tfrac{1}{2}xy\right) + z^2.$$

Methods 1 and 2 give two different expressions. However, these two expressions must be equivalent because they represent the same area. Therefore,

$$\text{area from Method 1} = \text{area from Method 2}$$
$$(x + y)^2 = 4(\tfrac{1}{2}xy) + z^2.$$

The brackets can be expanded and simplified. Therefore,

$$x^2 + y^2 + 2xy = 2xy + z^2.$$

The $2xy$ can be cancelled from both sides. So we have

$$x^2 + y^2 = z^2,$$

which is Pythagoras' theorem!

The argument is based on the fact that the area of the large square must be the same no matter what method is used to calculate it. We then logically derive two expressions for the same area, make them equivalent, and eventually the inevitable conclusion is that $x^2 + y^2 = z^2$, i.e. the square on the hypotenuse, z^2, is equal to the sum of the squares on the other two sides, $x^2 + y^2$.

This argument holds true for all right-angled triangles. The sides of the triangle in our argument are represented by x, y and z, and can therefore represent the sides of any right-angled triangle.

Appendix 2. Euclid's Proof that $\sqrt{2}$ is Irrational

Euclid's aim was to prove that $\sqrt{2}$ could not be written as a fraction. Because he was using proof by contradiction, the first step was to assume that the opposite was true, that is to say, that $\sqrt{2}$ could be written as some unknown fraction. This hypothetical fraction is represented by p/q, where p and q are two whole numbers.

Before embarking on the proof itself, all that is required is a basic understanding of some properties of fractions and even numbers.

(1) If you take any number and multiply it by 2, then the new number must be even. This is virtually the definition of an even number.

(2) If you know that the square of a number is even, then the number itself must also be even.

(3) Finally, fractions can be simplified: $\frac{16}{24}$ is the same as $\frac{8}{12}$; just divide the top and bottom of $\frac{16}{24}$ by the common factor 2. Furthermore, $\frac{8}{12}$ is the same as $\frac{4}{6}$, and in turn $\frac{4}{6}$ is the same as $\frac{2}{3}$. However, $\frac{2}{3}$, cannot be simplified any further because 2 and 3 have no common factors. It is impossible to keep on simplifying a fraction forever.

Now, remember that Euclid believes that $\sqrt{2}$ cannot be written as a fraction. However, because he adopts the method of proof by contradiction, he works on the assumption that the fraction p/q does exist and then he explores the consequences of its existence:

$$\sqrt{2} = p/q.$$

If we square both sides, then

$$2 = p^2/q^2.$$

This equation can easily be rearranged to give

$$2q^2 = p^2.$$

Now from point (1) we know that p^2 must be even. Furthermore, from point (2) we know p itself must also be even. But if p is even, then it can be written as $2m$, where m is some other whole number. This follows from point (1). Plug this back into the equation and we get

$$2q^2 = (2m)^2 = 4m^2.$$

Divide both sides by 2, and we get

$$q^2 = 2m^2.$$

But by the same arguments we used before, we know that q^2 must be even, and so q itself must also be even. If this is the case, then q can be written as $2n$, where n is some other whole number. If we go back to the beginning, then

$$\sqrt{2} = p/q = 2m/2n.$$

The $2m/2n$ can be simplified by dividing top and bottom by 2, and we get

$$\sqrt{2} = m/n.$$

We now have a fraction m/n, which is simpler than p/q.

However, we now find ourselves in a position whereby we can repeat exactly the same process on m/n, and at the end of it we will generate an even simpler fraction, say g/h. This fraction can then be put through the mill again, and the new fraction, say e/f, will be simpler still. We can put this through the mill again, and repeat the process over and over again, with no end. But we know from point (3) that fractions cannot be simplified forever. There must always be a simplest fraction, but our original hypothetical fraction p/q does not seem to obey this rule. Therefore, we can justifiably say that we have reached a contradiction. If $\sqrt{2}$ could be written as a fraction the consequence would be absurd, and so it is true to say that $\sqrt{2}$ cannot be written as a fraction. Therefore $\sqrt{2}$ is an irrational number.

Appendix 3. The Riddle of Diophantus' Age

Let us call the length of Diophantus' life L. From the riddle we have a complete account of Diophantus' life which is as follows:

> 1/6 of his life, $L/6$, was spent as a boy,
>
> $L/12$ was spent as a youth,
>
> $L/7$ was then spent prior to marriage,
>
> 5 years later a son was born,
>
> $L/2$ was the life span of the son,
>
> 4 years were spent in grief before he died.

The length of Diophantus' life is the sum of the above:

$$L = \frac{L}{6} + \frac{L}{12} + \frac{L}{7} + 5 + \frac{L}{2} + 4.$$

We can then simplify the equation as follows:

$$L = \frac{25}{28} L + 9,$$

$$\frac{3}{28} L = 9,$$

$$L = \frac{28}{3} \times 9 = 84.$$

Diophantus died at the age of 84 years.

Appendix 4. Bachet's Weighing Problem

In order to weigh any whole number of kilograms from 1 to 40 most people will suggest that six weights are required: 1, 2, 4, 8, 16, 32 kg. In this way, all the weights can easily be achieved by placing the following combinations in one pan:

$$1 \text{ kg} = 1,$$
$$2 \text{ kg} = 2,$$
$$3 \text{ kg} = 2 + 1,$$
$$4 \text{ kg} = 4,$$
$$5 \text{ kg} = 4 + 1,$$
$$\vdots$$
$$40 \text{ kg} = 32 + 8.$$

However, by placing weights in both pans, such that weights are also allowed to sit alongside the object being weighed, Bachet could complete the task with only four weights: 1, 3, 9, 27 kg. A weight placed in the same pan as the object being weighed effectively assumes a negative value. Thus, the weights can be achieved as follows:

$$1 \text{ kg} = 1,$$
$$2 \text{ kg} = 3 - 1,$$
$$3 \text{ kg} = 3,$$
$$4 \text{ kg} = 3 + 1,$$
$$5 \text{ kg} = 9 - 3 - 1,$$
$$\vdots$$
$$40 \text{ kg} = 27 + 9 + 3 + 1.$$

Appendix 5. Euclid's Proof That There Are an Infinite Number of Pythagorean Triples

A Pythagorean triple is a set of three whole numbers, such that one number squared added to another number squared equals the third number squared. Euclid could prove that there are an infinite number of such Pythagorean triples.

Euclid's proof begins with the observation that the difference between successive square numbers is always an odd number:

$$1^2 \quad 2^2 \quad 3^2 \quad 4^2 \quad 5^2 \quad 6^2 \quad 7^2 \quad 8^2 \quad 9^2 \quad 10^2 \ldots$$

$$1 \quad 4 \quad 9 \quad 16 \quad 25 \quad 36 \quad 49 \quad 64 \quad 81 \quad 100 \ldots$$

$$\backslash \; / \quad \backslash \; / \quad \backslash \; / \quad \backslash \; / \quad \backslash \; / \quad \backslash \; / \quad \backslash \; / \quad \backslash \; / \quad \backslash \; /$$

$$3 \quad 5 \quad 7 \quad 9 \quad 11 \quad 13 \quad 15 \quad 17 \quad 19 \quad \ldots$$

Every single one of the infinity of odd numbers can be added to a particular square number to make another square number. A fraction of these odd numbers are themselves square, but a fraction of infinity is also infinite.

Therefore there are also an infinity of odd square numbers which can be added to one square to make another square number. In other words there must be an infinite number of Pythagorean triples.

Appendix 6. Proof of the Dot Conjecture

The dot conjecture states that it is impossible to draw a dot diagram such that every line has at least three dots on it.

Although this proof requires a minimal amount of mathematics, it does rely on some geometrical gymnastics, and so I would recommend careful contemplation of each step.

First consider an arbitrary pattern of dots and the lines which connect every dot to every other one. Then, for each dot, work out its distance to the closest line, excluding any lines which go through it. Thereby identify which of all the dots is closest to a line.

Below is a close-up of such a dot D which is closest to a line L. The distance between the dot and the line is shown as a dashed line and this distance is smaller than any other distance between any other line and a dot.

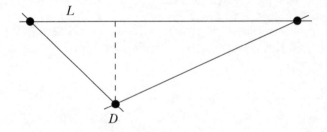

It is now possible to show that line L will always have only two dots on it and that therefore the conjecture is true, i.e. it is impossible to draw a diagram such that every line has three dots on it.

To show that line *L* must have two dots, we consider what would happen if it had a third dot. If the third dot, D_A, existed outside the two dots originally shown, then the distance shown as a dotted line would be shorter than the dashed line which was supposed to be the shortest distance between a dot and a line. Therefore dot D_A cannot exist.

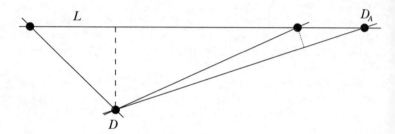

Similarly, if the third dot, D_B, exists between the two dots originally shown, then once again the distance shown as a dotted line would be shorter than the dashed line which was supposed to be the shortest distance between a dot and a line. Therefore dot D_B cannot exist either.

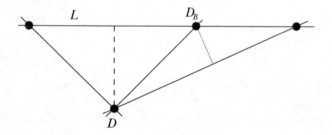

In summary, any configuration of dots must have a minimum distance between some dot and some line, and the line in question must have only two dots. Therefore for every configuration there will always be at least this one line with only two dots – the conjecture is true.

Appendix 7. Straying into Absurdity

The following is a classic demonstration of how easy it is to start off with a very simple statement and then within a few apparently straightforward and logical steps show that $2 = 1$.

First, let us begin with the innocuous statement

$$a = b.$$

Then multiply both sides by a, giving

$$a^2 = ab.$$

Then add $a^2 - 2ab$ to both sides:

$$a^2 + a^2 - 2ab = ab + a^2 - 2ab.$$

This can be simplified to

$$2(a^2 - ab) = a^2 - ab.$$

Finally, divide both sides by $a^2 - ab$, and we get

$$2 = 1.$$

The original statement appears to be, and is, completely harmless, but somewhere in the step-by-step manipulation of the equation there was a subtle but disastrous error which leads to the contradiction in the final statement.

In fact, the fatal mistake appears in the last step in which both sides are divided by $a^2 - ab$. We know from the original statement that $a = b$, and therefore dividing by $a^2 - ab$ is equivalent to dividing by zero.

Dividing anything by zero is a risky step because zero will go into any finite quantity an infinite number of times. By creating infinity on both sides we have effectively torn apart both halves of the equation and allowed a contradiction to creep into the argument.

This subtle error is typical of the sort of blunder which caught out many of the entrants for the Wolfskehl Prize.

Appendix 8. The Axioms of Arithmetic

The following axioms are all that are required as the foundation for the elaborate structure of arithmetic:

1. For any numbers m, n

$$m + n = n + m \quad \text{and} \quad mn = nm.$$

2. For any numbers m, n, k,

$$(m + n) + k = m + (n + k) \quad \text{and} \quad (mn)k = m(nk).$$

3. For any numbers m, n, k

$$m(n + k) = mn + mk.$$

4. There is a number 0 which has the property that, for any number n,

$$n + 0 = n.$$

5. There is a number 1 which has the property that, for any number n,

$$n \times 1 = n.$$

6. For every number n, there is another number k such that

$$n + k = 0.$$

7. For any numbers m, n, k,

$$\text{if } k \neq 0 \quad \text{and} \quad kn = km, \quad \text{then} \quad m = n.$$

From these axioms other rules can be proved. For example, by rigorously applying the axioms and assuming nothing else, we can rigorously prove the apparently obvious rule that

$$\text{if } m + k = n + k, \quad \text{then} \quad m = n$$

To begin with we state that

$$m + k = n + k.$$

Then by Axiom 6, let l be a number such that, $k + l = 0$, so

$$(m + k) + l = (n + k) + l.$$

Then, by Axiom 2,

$$m + (k + l) = n + (k + l).$$

Bearing in mind that $k + l = 0$, we know that

$$m + 0 = n + 0.$$

By applying Axiom 4, we can at last declare what we set out to prove:

$$m = n.$$

Appendix 9. Game Theory and the Truel

Let us examine Mr Black's options. First, Mr Black could aim at Mr Grey. If he is successful then the next shot will be taken by Mr White. Mr White has only one opponent left, Mr Black, and as Mr White is a perfect shot then Mr Black is a dead man.

A better option is for Mr Black to aim at Mr White. If he is successful then the next shot will be taken by Mr Grey. Mr Grey hits his target only two times out of three and so there is a chance that Mr Black will survive to fire back at Mr Grey and possibly win the truel.

It appears that the second option is the strategy which Mr Black should adopt. However, there is a third and even better option. Mr Black could aim into the air. Mr Grey has the next shot and he will aim at Mr White, because he is the more dangerous opponent. If Mr White survives then he will aim at Mr Grey because he is the more dangerous opponent. By aiming into the air, Mr Black is allowing Mr Grey to eliminate Mr White or vice versa.

This is Mr Black's best strategy. Eventually Mr Grey or Mr White will die and then Mr Black will aim at whoever survives. Mr Black has manipulated the situation so that, instead of having the first shot in a truel, he has first shot in a duel.

Appendix 10. An Example of Proof by Induction

Mathematicians find it useful to have neat formulae which give the sum of various lists of numbers. In this case the challenge is to find a formula which gives the sum of the first n counting numbers.

For example, the sum of just the first number is 1, the sum of the first two numbers is 3 (i.e. $1 + 2$), the sum of the first three numbers is 6 (i.e. $1 + 2 + 3$), the sum of the first four numbers is 10 (i.e. $1 + 2 + 3 + 4$), and so on.

A candidate formula which seems to describe this pattern is:

$$\text{Sum}(n) = \tfrac{1}{2}n(n + 1).$$

In other words if we want to find the sum of the first n numbers, then we simply enter that number into the formula above and work out the answer.

Proof by induction can prove that this formula works for every number up to infinity.

The first step is to show that the formula works for the first case, $n = 1$. This is fairly straightforward, because we know that the sum of just the first number is 1, and if we enter $n = 1$ into the candidate formula we get the correct result:

$$\text{Sum}(n) = \tfrac{1}{2}n(n + 1)$$

$$\text{Sum}(1) = \tfrac{1}{2} \times 1 \times (1 + 1)$$

$$\text{Sum}(1) = \tfrac{1}{2} \times 1 \times 2$$

$$\text{Sum}(1) = 1.$$

The first domino has been toppled.

The next step in proof by induction is to show that if the formula is true for any value n, then it must also be true for $n + 1$. If

$$\text{Sum}(n) = \tfrac{1}{2}n(n + 1),$$

then,

$$\text{Sum}(n + 1) = \text{Sum}(n) + (n + 1)$$

$$\text{Sum}(n + 1) = \tfrac{1}{2}n(n + 1) + (n + 1).$$

After rearranging and regrouping the terms on the right, we get

$$\text{Sum}(n + 1) = \tfrac{1}{2}(n + 1)\,[(n + 1) + 1].$$

What is important to note here is that the form of this new equation is exactly the same as the original equation except that every appearance of n has been replaced by $(n + 1)$.

In other words, if the formula is true for n, then it must also be true for $n + 1$. If one domino falls, it will always knock over the next one. The proof by induction is complete.

Suggestions for Further Reading

In researching this book I have relied on numerous books and articles. In addition to my main sources for each chapter, I have also listed other material which may be of interest to both the general reader and experts in the field. Where the title of the source does not indicate its relevance I have given a sentence or two describing its contents.

Chapter 1

The Last Problem, by E.T. Bell, 1990, Mathematical Association of America. A popular account of the origins of Fermat's Last Theorem.

Pythagoras – A Short Account of His Life and Philosophy, by Leslie Ralph, 1961, Krikos.

Pythagoras – A Life, by Peter Gorman, 1979, Routledge and Kegan Paul.

A History of Greek Mathematics, Vols. 1 and 2, by Sir Thomas Heath, 1981, Dover.

Mathematical Magic Show, by Martin Gardner, 1977, Knopf. A collection of mathematical puzzles and riddles.

River meandering as a self-organization process, by Hans-Henrik Støllum, *Science* **271** (1996), 1710–1713.

Chapter 2

The Mathematical Career of Pierre de Fermat, by Michael Mahoney, 1994, Princeton University Press. A detailed investigation into the life and work of Pierre de Fermat.

Archimedes' Revenge, by Paul Hoffman, 1988, Penguin. Fascinating tales which describe the joys and perils of mathematics.

Chapter 3

Men of Mathematics, by E.T. Bell, Simon and Schuster, 1937. Biographies of history's greatest mathematicians, including Euler, Fermat, Gauss, Cauchy and Kummer.

The periodical cicada problem, by Monte Lloyd and Henry S. Dybas, *Evolution* **20** (1966), 466–505.

Women in Mathematics, by Lynn M. Osen, 1994, MIT Press. A largely non-mathematical text containing the biographies of many of the foremost female mathematicians in history, including Sophie Germain.

Math Equals: Biographies of Women Mathematicians+Related Activities, by Teri Perl, 1978, Addison-Wesley.

Women in Science, by H. J. Mozans, 1913, D. Appleton and Co.

Sophie Germain, by Amy Dahan Dalmédico, *Scientific American,* December 1991. A short article describing the life and work of Sophie Germain.

Fermat's Last Theorem – A Genetic Introduction to Algebraic Number Theory, by Harold M. Edwards, 1977, Springer. A mathematical discussion of Fermat's Last Theorem, including detailed outlines of some of the early attempts at a proof.

Elementary Number Theory, by David Burton, 1980, Allyn & Bacon.

Various communications, by A. Cauchy, *C. R. Acad. Sci. Paris* **24** (1847), 407–416, 469–483.

Note au sujet de la demonstration du theoreme de Fermat, by G. Lamé, *C. R. Acad. Sci. Paris* **24** (1847), 352.

Extrait d'une lettre de M. Kummer à M. Liouville, by E.E. Kummer, *J. Math. Pures et Appl.* **12** (1847), 136. Reprinted in *Collected Papers,* Vol. I, edited by A. Weil, 1975, Springer.

A Number for Your Thoughts, by Malcolm E. Lines, 1986, Adam Hilger. Facts and speculations about numbers from Euclid to the latest computers, including a slightly more detailed description of the dot conjecture.

Chapter 4

3.1416 and All That, by P.J. Davis and W.G. Chinn. 1985, Birkhäuser. A series of stories about mathematicians and mathematics, including a chapter about Paul Wolfskehl.

The Penguin Dictionary of Curious and Interesting Numbers, by David Wells, 1986, Penguin.

The Penguin Dictionary of Curious and Interesting Puzzles, by David Wells, 1992, Penguin.

Sam Loyd and his Puzzles, by Sam Loyd (II), 1928, Barse and Co.

Mathematical Puzzles of Sam Loyd, by Sam Loyd, edited by Martin Gardner, 1959, Dover.

Riddles in Mathematics, by Eugene P. Northropp, 1944, Van Nostrand.

The Picturegoers, by David Lodge, 1993, Penguin.

13 Lectures on Fermat's Last Theorem, by Paulo Ribenboim, 1980, Springer. An account of Fermat's Last Theorem, written prior to the work of Andrew Wiles, aimed at graduate students.

Mathematics: The Science of Patterns, by Keith Devlin, 1994, Scientific American Library. A beautifully illustrated book which conveys the concepts of mathematics through striking images.

Mathematics: The New Golden Age, by Keith Devlin, 1990, Penguin. A popular and detailed overview of modern mathematics, including a discussion on the axioms of mathematics.

The Concepts of Modern Mathematics, by Ian Stewart, 1995, Penguin.

Principia Mathematica, by Betrand Russell and Alfred North Whitehead, 3 vols , 1910, 1912, 1913, Cambridge University Press.

Kurt Gödel, by G. Kreisel, Biographical Memoirs of the Fellows of the Royal Society, 1980.

A Mathematician's Apology, by G.H. Hardy, 1940, Cambridge University Press. One of the great figures of twentieth-century mathematics gives a personal account of what motivates him and other mathematicians.

Alan Turing: The Enigma of Intelligence, by Andrew Hodges, 1983, Unwin Paperbacks. An account of the life of Alan Turing, including his contribution to breaking the Enigma code.

Chapter 5

Yutaka Taniyama and his time, by Goro Shimura, *Bulletin of the London Mathematical Society* **21** (1989), 186–196. A very personal account of the life and work of Yutaka Taniyama.

Links between stable elliptic curves and certain diophantine equations, by Gerhard Frey, *Ann. Univ. Sarav. Math. Ser.* **1** (1986), 1–40. The crucial paper which suggested a link between the Taniyama–Shimura conjecture and Fermat's Last Theorem.

Chapter 6

Genius and Biographers: the Fictionalization of Evariste Galois, by T. Rothman, *Amer. Math. Monthly* **89** (1982), 84–106. Contains a detailed list of the historical sources behind Galois's biographies, and discusses the validity of the various interpretations.

La vie d'Evariste Galois, by Paul Depuy, *Annales Scientifiques de l'Ecole Normale Supérieure* **13** (1896), 197–266.

Mes Memoirs, by Alexandre Dumas, 1967, Editions Gallimard.

Notes on Fermat's Last Theorem, by Alf van der Poorten, 1996, Wiley. A technical description of Wiles's proof aimed at mathematics undergraduates and above.

Chapter 7

An elementary introduction to the Langlands programme, by Stephen
  Gelbart, *Bulletin of the American Mathematical Society* **10** (1984), 177–219.
  A technical explanation of the Langlands programme aimed at
  mathematical researchers.
Modular elliptic curves and Fermat's Last Theorem, by Andrew Wiles,
  Annals of Mathematics **142** (1995), 443–551. This paper includes the bulk
  of Wiles's proof of the Taniyama–Shimura conjecture and Fermat's
  Last Theorem.
Ring-theoretic properties of certain Hecke algebras, by Richard Taylor
  and Andrew Wiles, *Annals of Mathematics* **142** (1995), 553–572. This
  paper describes the mathematics which was used to overcome the flaws
  in Wiles's 1993 proof.

Chapter 8

How to succeed in stacking, by Ian Stewart, *New Scientist*, 13 July 1991, pp.
  29–32.
The death of proof, by John Horgan, *Scientific American*, October 1993, pp.
  74–82.
The solution of the four-color-map problem, by Kenneth Appel and
  Wolfgang Haken, *Scientific American*, October 1977, pp. 108–21.
The Four-Color Problem: Assaults and Conquest, by T. L. Saaty and P.C.
  Kainen, McGraw-Hill, 1977.
The Mathematical Experience, by P. J. Davis and R. Hersh, 1990, Penguin.

Picture Credits

Index